THE SACRED CHAIN

THE
SACRED
CHAIN

HOW UNDERSTANDING EVOLUTION
LEADS TO DEEPER FAITH

JIM STUMP

HarperOne

An Imprint of HarperCollins*Publishers*

HarperCollins books may be purchased for educational, business, or sales promotional use. For information, please email the Special Markets Department at SPsales@harpercollins.com.

FIRST HARPERONE PAPERBACK EDITION PUBLISHED IN 2025

Illustrations © Sloan Stump
Hexagon art © natrot/stock.adobe.com

Library of Congress Cataloging-in-Publication Data is available upon request.

ISBN 978-0-06-335095-3

25 26 27 28 29 LBC 5 4 3 2 1

For Finley and Auden

CONTENTS

Part III: Species

Part IV: Soul

Part V: Pain

INTRODUCTION

I sat by myself at one end of the boardroom, fidgeting with a few notes on the table in front of me. At the other end were about ten older men in suits and ties peering over their tables, arms crossed. It wouldn't have been difficult for someone just walking in to determine which end of the room held the power.

After my brief opening statement, the rest of our time was set aside for "discussion." The two-hour meeting felt more like an inquisition. The first questioner hardly let the silence settle after I spoke: "Jim, I went to school with your father. We even went on a mission trip to Mexico together. I've known you since you were a kid. What happened to you?"

I've replayed this scene in my head a hundred times, varying what I say in an attempt to get my accusers to stand up, shake my hand, and say, "Oh, *now* we see. That makes sense. Sorry for the trouble." But every time it ends the same way: I have to give up my position as a tenured professor of philosophy and leave the college I've served for seventeen years.

My crime? Believing what 99 percent of those who have a PhD in biology or medicine believe: that human beings evolved over time and share common ancestors with every other life-form on the planet. But this was a small Christian college, one of the places where evolutionary theory is deemed incompatible with Christian beliefs. And not just incompatible—evolution is considered dangerous. These men believed that hearing anything positive about evolution would make students doubt the Bible: if you can't believe the creation account in the very first chapter of the Bible, their thinking goes, then why believe any of it?

I do not believe faith is so fragile. I'd already shared with the panel examples of the ways faith and science can not only coexist but actually strengthen each other. But the rhetoric of prominent creationist groups—groups whose rigid interpretations are rooted in a creation science movement less than a hundred years old—so unsettled the college leadership that, after questioning me, they changed the official statement of beliefs that faculty have to sign each year. Not only was I forced out of my job, but my students would no longer be able to learn from a professor who openly affirmed the scientific consensus on evolution.

To be fair, I wasn't technically fired. I could have stayed if I agreed to the new rules. There are still a few other faculty members there who would admit behind closed doors that they accept evolution. But they can't teach it as true, like they do photosynthesis or germ theory. They can't publish scholarly work that defends it. And they *certainly* can't have leadership positions in organizations that advocate for evolution—even if that advocacy comes from a Christian viewpoint.

That last point was the kicker for me. I had been working with one such organization, BioLogos, for a couple of years by that time. BioLogos was founded by Dr. Francis Collins—the scientist who led the international Human Genome Project and later served as director of the National Institutes of Health under Presidents Obama, Trump, and Biden. In 2006, Collins published a bestselling book, *The Language of God*, about his adult conversion to the Christian faith and how he made sense of his religious beliefs alongside the science he practiced. BioLogos was founded to spread that message of harmony to others: you don't have to choose between the well-established science of our day—including evolution—and authentic Christian faith. That news clearly has not gotten out to everyone yet.

I had taken on a part-time position at BioLogos in 2013 with full approval from the college administration. When I first brought up the role, the president was not only supportive but enthusiastic. He wanted the college to be part of these important conversations at the intersection of faith and culture. But he was new to our community, having just been hired from the outside. He understood that evolution could be a controversial topic, but he was confident that we could handle it well. Maybe too confident.

Some of my friends and colleagues had wondered how this arrangement would go over with the broader constituents, but most of them thought I'd be safe because I was an insider. I did my undergraduate degree at that same college, and one semester when I was there, eight of my first cousins were enrolled there too! We got our picture on the front page of the local newspaper. My parents both went to college there, along with a bunch of my aunts and uncles. There are even two buildings on campus named after my ancestors.

Yes, my roots ran deep in that institution. I wasn't coming in from the outside to stir up trouble. I wasn't even trying to stir up trouble from the inside. I was simply tired of hearing that former students had given up their faith over a supposed conflict with science. They had been given a view of science—sometimes intentionally, sometimes simply absorbed from their surroundings—that just didn't hold up when they got out into the real world. These were not a few isolated cases. Science in general, and evolution in particular, have been responsible for the deconstruction of religious faith for many people. But I had learned from my own experience that a better understanding of evolution could actually lead to a deeper and more authentic faith.

So I started talking more frequently about the relationship between science and faith. I tried to show students ways of understanding the Bible that don't require us to dismiss the work of experts in the sciences. I did that with the explicit approval of the administration. But word got out in wider circles that the college had an "evolutionist" on the faculty. And when I started working with BioLogos, my views were not limited to classrooms or conversations in my office. They were published on a website that was becoming a little too popular for me to stay out of the public eye. Influential people in the college community were upset, and the wheels were set in motion that led to my departure. My picture appeared on the front page of the local newspaper again, this time with the headline, "Professor Resigns After College Stance on Human Origins." To this day, I don't know how the story got out.

It was a very hard year for my family and me. But with some distance, I now see that trying time as a net positive. It might not have been a good thing in and of itself, but I can affirm that God used it for good. I began working full-time at BioLogos and discovered a rewarding second career there. I know personally half a dozen other faculty members who have had to leave Christian colleges over their affirmation of mainstream science, and I've heard of many more. Most of them did not have a place to land like I did, where they could earn a living doing what they're passionate about. I was the fortunate one. But our collective experience underscores the need for a book like this.

A word about the title: as we were preparing the proposal to send to publishers, Gail, my literary agent, said, "We need to settle on a title and subtitle to submit." I had been using a working title we weren't in love with for various reasons. One of her colleagues, Dara, was part of our conversation, and she immediately produced a list of possible titles.

Gail said, "Wow, did you just come up with those?"

Dara responded, "No, I just asked ChatGPT to suggest some titles."

These were the days when ChatGPT was new to the headlines as the latest artificial intelligence bot that users could interact with. The title that stood out to us from its list was *The Sacred Chain*.

The word *chain* evokes the DNA molecule, and calling it "sacred" makes us think God might have had something to do with it. There is also a chain of religious doctrine and tradition that links us to the past. I usually get nervous when we combine scientific and theological claims too intimately. For example, I don't think it's properly scientific to say "and then a miracle happened" nor properly theological to think science can explain God. But the title of a book about science and faith might be the place for an exception to that rule. It's tantalizing to suggest that there could be more going on in the evolution of our bodies than science alone can describe and that science places parameters on the development of our spiritual lives.

With a little more poking around the internet, I discovered that a chain is also a unit of measurement used predominantly by surveyors. A chain is 66 feet, which means there are exactly 80 chains in that otherwise curious total of 5,280 feet in a mile. And an acre is exactly equal to ten square chains. So here's another (admittedly looser) connotation of my title: a book that surveys a vast swath of our species' past and finds important and sacred features of that landscape.

Furthermore—and even more loosely (though this is what really sold me on the title)—the chain unit of measurement is what is used to set the distance between stumps on a cricket pitch. For the majority of Americans who don't know what those words mean, a pitch is what the Brits call a field on which sports are played, and cricket is one of those quintessentially British sports during which everyone stops to have tea. I learned the basic rules of the game when my family took a trip to Oxford one summer for a fellowship I had there, and we were renting (I should say "letting") an apartment (I mean a "flat") from a family. We met them for an afternoon and decided to let the kids run off some energy in the backyard (er, "garden").

They had a cricket set and got a pretty big laugh about our last name and our three kids, because the stumps in cricket are the three poles that are driven into the ground and on top of which the bales are balanced. (We attempted to line up the kids and balance a stick on top of them, but their significant height differences made this impossible.) The batter stands in front of this contraption, collectively called a "wicket," and tries to protect the bales from being knocked off the stumps by the bowler on the other team who throws a ball at them. Exactly one chain away is another set of stumps the batter runs to when the ball is hit.

It seemed just too good that ChatGPT would come up with a title that had a connection to my last name. So I asked Dara if she had given it any information about me as the author. She had not. It turns out that there are some coincidences—both in AI and in the history of life!

We still needed a subtitle. After much discussion with the publishing team, we settled on *How Understanding Evolution Leads to Deeper Faith*. That's really the point of the book. Too many loud voices proclaim that science has shown religion to be silly at best and dangerous at worst. And too many people feel that the only alternative to this view is to join with the loud voices on the other side of the conflict who reject the findings of modern science in order to preserve their understanding of religious faith. I think there is a better way—one that not only preserves faith in the face of modern science but, through significant engagement and dialogue with science, leads to a deeper, stronger, and more relevant faith in today's world. At least, that's what it did for me.

This book shows how I worked through the challenge of reconciling religious truths with scientific ones and found new and more profound insights about God and humanity along the way. As I tested out approaches to bringing these two ways of seeing the world into alignment, it felt a bit like adjusting a pair of binoculars this way and that until the complete image came into focus. I gained both clarity and perspective. But some areas remained blurry longer than others.

The book is organized around my explorations of these initially blurry areas, which presented challenges to seeing evolution and Christian faith as one coherent picture.

The chapters in part I are about the challenge of the Bible—why does evolution pose such a threat to so many American Christians' view of the Bible, and how might we think differently about it? I give some statistics showing that the United States is a significant outlier when it comes to evolution acceptance, and I tease out some consequences of how creationists use the Bible. Finally, I look to C. S. Lewis for a better way of understanding scripture. This not only helped me reconcile modern science and the Bible but also gave me tools for tackling the other challenges.

Part II begins with a fun exercise to help us grasp the incredibly vast stretch of time our universe has existed. Humans are just a tiny blip on this scale. What does this say about God's priorities? I describe my visits to some very old "calendars" and explain other methods scientists have developed to keep track of time. This understanding of time prompts me to go back to the origin stories in the Bible and read them less like newspaper accounts and find richer insights about the human condition. And I benefit from the insights of another Brit, G. K. Chesterton, and from Indian author Arundhati Roy. They helped me see God less as an engineer and more as an artist who delights in small and seemingly insignificant details.

The challenge of part III is the concept of a species. I had been led by my theological tradition and by the hard-wiring of our brains to think of species as fixed and unchanging. But evolution reveals that the boundaries between them are fuzzy and constantly changing. What does that mean for humans, who according to the Bible are created in the image of God? Is that designation just for *Homo sapiens*? What about the cast of other humanlike characters that scientists have discovered and that the biblical authors knew nothing about, like Neanderthals? Do we really differ *in kind* from all other creatures, or is it only a difference *of degree*? I unsuccessfully tried to visit a couple of ancient archaeological sites but then was successful in visiting ancient cave paintings and old-growth redwood trees. These failures and successes gave me fresh insights about our species and a better understanding of our place in creation.

Part IV considers the challenge of how we can understand the soul if the human body is the product of evolution. Saint Gregory of Nyssa (along with an Eastern Orthodox priest whose church in San Diego supposedly has a relic of Saint Gregory) helped me sort out the importance of our physical bodies for the capacities of our minds and souls. Walking upright on two legs had a surprisingly significant effect on the kind of creature our ancestors became over the last several million years, including the game-changing ability to speak. I describe some of the fossil discoveries that have shed light on walking and talking, and I consider the testimony of Helen Keller about the different world that opens up to us when we have language. We see how that world allows us to talk about souls.

The challenge in part V is the hardest. I won't pretend to neatly solve the age-old problem of why there is pain and suffering in a good world created by God. But I do believe that bringing this problem into conversation with evolutionary science points to a more satisfying solution than that of my theological tradition. The evolutionary development of our capacity to love takes center stage here, and I use the ideas of French philosopher and mystic Simone Weil to contemplate how God might use pain and suffering for important and good purposes.

In the conclusion, I reflect on what the role of suffering in our past means for our present and future. This involves speculation about the ultimate fate of the universe and ourselves from the perspectives of science and faith. But I suggest that there are legitimate grounds for hope—not just wishful thinking—that the present order of things is not the end.

This book should not be understood primarily as a "Christian" book, one relevant only for followers of Jesus. There is no denying that I write from the Christian tradition; I was born into it, and it has shaped my thinking and imagination. And no matter how embarrassed I am by the words and deeds of some others who identify as Christian today, I continue to belong to a generous Christian community that attempts to live and love according to the example of Christ. But I am not arguing that mine is the only authentic articulation of faith, and I hope this book will appeal across traditional lines and be relevant to anyone who believes (or simply hopes) that there is a personal Being at the core of the universe who loves us.

My engagement with science has led me to a deeper, more authentic faith. This book tells the story of how that happened through lots of conversations with interesting people and visits to remarkable places. All of the personal anecdotes I tell are true, but this is not a strictly chronological account of my life. In real life, working through ideas like these happens simultaneously on many fronts, rather than sequentially. But my goal here is conceptual clarity, so I've taken conversations and episodes from my life and packaged them so they might be most easily understood. In that sense, my story is an everyman story—a kind of archetype for others who have wondered how science and religion might fit together.

As such, reading this book about my journey might be the start of one's own journey for some readers. Perhaps you will find yourself considering ideas your religious community has told you to stay away from; you might come to see that evolutionary science can help us better understand a God who intentionally created human beings. Or perhaps you will begin to wonder if there is more to our story than what science can tell; maybe you'll start to think that religion, as flawed and shot through with human ineptitude as it is, might really be pointing at something beyond what science can explain. Maybe some minds will be changed as a result.

I know this book will probably not convince everyone. But I'm hopeful that many people will find in these pages a better (and more accurate!) account of the whole scope of our humanity—that they will follow the sacred chain that links our biological past with our present capacity for morality and even the hope for future immortality.

PART I

BIBLE

COMMUNITIES OF ORIGIN

My journey to understanding science and faith began where everyone's does: in the community I was born into. I come from a very religious family. Every week when I was growing up, we went to Sunday school, Sunday morning worship, Sunday evening worship, and Wednesday evening services. On days we weren't in church, we almost always had family devotions—a time for group Bible reading and prayer. By the time I was a teenager, I regularly had my own private time for Bible reading and study, and I participated on the Bible quiz team at my church, which involved memorizing entire books of the Bible. I went to Bible camps in the summer and was only allowed to listen to Christian music, where the lyrics were mostly taken from the Bible.

That might sound like a stern, puritanical upbringing, but I don't remember it that way. We took the Bible very seriously, but there was also a lot of fun. And we weren't cloistered away from the rest of society. My family may have been on the more devout end of the spectrum, but we weren't that different from others in our local community.

My parents were both educators in the public school system, and those were the schools I went to. Public schools in small Midwestern towns were not on the cutting edge of science education in their day. I wish I could remember when I first encountered the scientific theory of evolution, but I don't think it was ever a topic in any science class I had all the way through high school. It was also never addressed in my Science Education major in college. Evolution just wasn't a topic in the science curriculum of my local community.

By the time my own kids were in junior high around 2010, in a very large public school system (also in the Midwest), schools had at least begun to teach evolution. But their eighth-grade biology teacher introduced the unit by saying: "Well, they force me to teach you about evolution, but I want you to know that I don't think it is real."

When I was confirming this story with my kids to make sure I wasn't misremembering the situation, they said, "Yes, and the same thing happened in social studies class."

"What?!" I responded. "Why was your social studies teacher

in junior high trying to talk you out of accepting evolutionary theory?" Evidently, there was a unit on the early development of human beings that noted the different species of "humans" that had evolved from common ancestors. There too the teacher just stated matter-of-factly that he didn't think it was true. I asked my kids if they remember any reasons that were given for these teachers' rejection of a well-established scientific theory.

"No, they didn't try to argue against it with scientific reasons. They just said they didn't believe evolution was true because of their faith and belief in the Bible, and then moved on to teach what the textbook said."

This was the local community I was part of. Of course, we're all part of lots of different communities to varying degrees: our neighborhood and city, work and social circles, even state and national communities. But overlying all of these for me was involvement in the religious community I had inherited and chose to continue participating in. Central to that community, and to the way we understood and interacted with all other communities, was the Bible.

But it was not just the Bible, as though anyone who finds wisdom and truth in the Bible would have the same problems with science. Instead, I think the problem stems from a veneration of the Bible, rather than a veneration of what the Bible bears witness to. That's a subtle distinction and needs a little further explanation.

It is easier to find clarity and certainty if you focus on the literal or plain meanings of the words in the Bible. I used to hear the slogan "The Bible says it, and that settles it" as though all the answers are there, and no further discussion or debate is required. That attitude treats the Bible as an authoritative answer book. For example, the Bible says God created the heavens and the Earth in six days, and Noah built an ark that housed two of every kind of animal. It's very straightforward to take these as answers to the questions "How old is the Earth?" and "Where did all our present-day animals come from?" It is a messier process and open to lots of debate if we understand these as culturally generated statements that might be pointing toward deeper, more universal truths.

If the words of the Bible are open to being interpreted in different ways, there would be endless debate, and we might never know what they really mean. Better to go with the "plain reading" of the Bible and leave it at that. And what of the fact that such a reading obviously contradicts modern science? Well, so much the worse for modern science.

This was the view of the Bible I had inherited, and as I progressed in my studies and eventually in my career, I wasn't sure what to do with it given what I had learned about evolution. I was comfortable with ambiguity and uncertainty, but bending the plain meanings of the words in scripture far enough to be compatible with science would cause me to run seriously afoul of my tradition.

For those not acquainted with this community, it could be tempting to think that people who hold a literal view of the Bible are uneducated or unintelligent. But that's not quite right. Education is a factor, as we'll see later, but I know some highly educated and intelligent Christians who hold to a version of the view that finds tension between the Bible and science. They apply their intellects to resolve those tensions in very creative ways and have a big influence on conservative communities.

For someone who is a product and a casualty of a community like this, it is tempting to ridicule and make fun of the community and its members. That's not my intention here. I want to understand them and how their views could depart so significantly from the findings of modern science, while at the same time pointing to what I believe is a better way of finding harmony between science and faith. I'll move from personal anecdotes to more general statistics.

There are several national surveys that look at what people believe about evolution. One of the more famous ones was published in 2006 by the prestigious journal *Science*. Researchers compared the acceptance of evolution in America and in thirty-three countries in Europe. The only country that had a lower public acceptance of evolution than the United States was Turkey.[1] These kinds of comparative studies can be tricky, though, because it wasn't the same polling group conducting the surveys in each country, and the questions weren't asked exactly the same way. Still, it's pretty remarkable how low America is on that list.

Comparing the trend in America over time is a little easier because the same question has been asked every few years since 1985 to a sample of adults.[2] They were asked whether they accept, reject, or don't know about the following statement:

Human beings, as we know them, developed from earlier species of animals.

The percentage of people answering "accept" was very close to the percentage of people answering "reject" from 1985 to 2007 (see graph below). These were around 40 percent each, with the "accepts" staying just barely ahead for most of that time. When the question was asked again in 2012, the "accept" response jumped substantially to over 45 percent, and then it rose again over the next few years to a high of 57 percent.

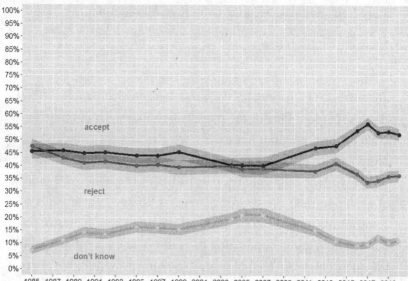

There are probably several reasons for this, but I'd like to think my organization, BioLogos, has had something to do with that change. This data tracks pretty closely with our founding and subsequent development of resources aimed specifically at showing that evolution and Christian faith can coexist. But correlation is not causation, and all I have is anecdotes to justify my hunch. Back to statistics.

A different survey conducted by the Pew Research Center in 2015[3] asked a representative sample of adults a slightly different question:

Thinking about evolution, which comes closer to your view:

Humans and other living things have evolved over time.

Humans and other living things have existed in their present form since the beginning of time.

The results in this survey were a little more encouraging. They found that 65 percent of US adults said that humans and other living things have evolved over time, and only 31 percent said they have existed in their present form since the beginning of time. One reason this question received a higher level of evolution acceptance could be that it asked "which comes closer to your view" without offering an "undecided" option.

Pew asked the same question again in 2019 and got nearly identical results for US adults (64 percent to 32 percent). But this time, they asked the same question to adults in nineteen other countries with "sizable or growing investments in scientific and technological development."[4] The US isn't at the bottom of this list for public acceptance of evolution, but it's still pretty close: we were fifteenth out of twenty.

I wanted to understand why Americans are so resistant to the science of evolution, so I dug deeper into these survey results.

Not surprisingly, when you look at how Christians respond to the question versus the religiously unaffiliated, Christians are less likely to accept evolution in all twenty countries. But the gap between Christians and the religiously unaffiliated was larger in the

US than in any other country. For example, in Spain, there is only a 7-point difference between Christians and the unaffiliated accepting evolution (86 percent to 93 percent); in Germany, the difference is only 4 points (81 percent to 85 percent); in Canada and the United Kingdom, that gap jumps to 23 percentage points. But in the US, a whopping 35 points separates acceptance of evolution between Christians (54 percent) and the unaffiliated (89 percent).

Also consistent across all twenty countries was the finding that people with more education are more likely to accept evolution. But again, that gap is largest in the US. People here who completed college with three or more courses in science accepted evolution at 13 percentage points higher (77 percent) than those who completed college with two or fewer courses in science (64 percent). That gap in the UK was only 6; in Canada, 5; and in Spain, 2.

It shouldn't shock us that religion and education affect one's acceptance of evolution. But why are those effects so much stronger in the US than elsewhere?

I host a podcast for BioLogos called *Language of God*—named after the book by Francis Collins (who also happens to be our most frequent guest on the show). Most of the episodes are conversations with people about science and religion, and whenever I talk to someone from another country, I ask about the difference in religious people accepting evolution where they're from compared to the US. Many of my guests think the difference is pretty remarkable, saying something like, "It just isn't an issue among church people where I'm from." When I ask why that is, they often say, "Well, you had the Scopes trial."

The Scopes trial happened in Tennessee in 1925 and was a national phenomenon. What the O. J. Simpson trial was for the age of television, and what the trial of George Floyd's murderer was for the age of the internet, the Scopes trial was for the age of newspapers. Everyone was paying attention to it. Officially, John Scopes was prosecuted for teaching evolution in violation of a Tennessee law that had recently been passed. But actually the trial was just the stage for a high-profile debate between Clarence Darrow and William Jennings Bryan.

Darrow was the defense lawyer for Scopes, and he knew his client was guilty. But he hoped the trial would garner the public's attention and expose the prosecution and its supporters as unscientifically minded, thereby getting the law overturned. He said, "We have the purpose of preventing bigots and ignoramuses from controlling the education of the United States."[5]

William Jennings Bryan had run for president of the US three times, and he used his celebrity status to campaign around the country in favor of anti-evolution laws. He allowed himself to be questioned on the stand by Darrow about his views on the Bible, and it became a national spectacle that forced people to choose sides. Unfortunately, the only sides available were among those who accepted modern science and those who trusted their Bibles.

But there is more to the story of the American rejection of evolution. I interviewed historian and theologian Stan Rosenberg about this for the podcast.[6] Stan was born in the US but has worked in England for more than twenty years. When I asked him about the difference between acceptance of evolution in the US compared to Europe, he said, "I think one has to take account of the fact that *On the Origin of Species* is published in 1859, and so it is coming to bear in the US—it is being transmitted to the US—just at the time of the US Civil War. And war changes and shapes and distorts a culture." Furthermore, he noted that evolution and its positing of continual gradual change is not a very intuitive idea. We seem hardwired to think about things persisting as they are.

So right at the time of a very deep division in our country, there was a new, counterintuitive idea from the sciences. Our culture, distorted by the Civil War, exaggerated cultural differences and bundled them together in ways that didn't necessarily have to be. Science ended up on one side, the Bible on the other. And when the Scopes trial became a national phenomenon, people had to pick which side they were going to identify with. In that sense, instead of being the primary cause of division, the Scopes trial was more like an occasion for the deeply settled division in the country to bubble to the surface again.

In the next chapter, I'll describe my experience with the more recent manifestation of this division.

CREATIONISM

When I was a college student (at that same school I had to leave), my major was called Science Education. But just for fun, I took quite a few elective courses in religion, the Bible, and what was called "apologetics" (which is not apologizing for faith but providing a rational defense of faith). I found the professors in those areas to be engaging and interesting. One of them was a creationist crusader, and almost everything I learned about evolution at the time came from him.

Creationist is an unfortunate term because almost everyone who believes in God believes that God is a creator in some sense. But especially since the 1960s and the publication of the book *The Genesis Flood* by John Whitcomb and Henry Morris, the terms *creationist* and *creationism* have taken on a more specific meaning in the science and religion community. The terms are typically used for those who don't accept that humans and other living things have evolved, but these creationists come in two varieties. Some people and organizations identify as "old Earth creationists"; they don't accept evolution (especially of humans), but they generally accept the physical sciences, like geology, physics, and cosmology, from which we get the accepted dating of Earth and the universe as billions of years old. Then there are the "young Earth creationists," who reject not only the evolutionary conclusions of the life sciences but also the dating conclusions of the physical sciences. Young Earth creationists typically believe that the Earth and the universe are only about 6,000 to 10,000 years old because, as we'll see later, that's the range of dates you find when taking the ages and genealogies in the Bible literally, going all the way back to Adam and Eve.

The Genesis Flood is a manifesto on how all the apparent scientific arguments for the ancient age of the Earth can be explained away by the supposedly biblical fact that there was a massive flood in the days of Noah that covered the entire planet. The book was enormously popular (it has sold hundreds of thousands of copies) and launched a whole generation of crusaders who thought that the authority of the Bible depends on modern science being wrong about evolution and the age of the Earth.

My young Earth creationist professor kept in touch with me throughout graduate school, attempting to keep me on the straight and (very) narrow path he deemed to be an acceptable form of Christian faith. At one point, he bought me a brand-new copy of *The Genesis Flood* with a short note written inside the front cover that read, "Jim, keep defending the faith non-fideistically. Exodus 20:11, Jude 3."

Let's unpack that a bit because it helps to understand this community. Here's the text of the Bible verses he cited:

> *Exodus 20:11: "For in six days the Lord made heaven and earth, the sea, and all that is in them."*

> *Jude 3: "I find it necessary to write and appeal to you to contend for the faith."*

The Exodus verse is a common one for creationists to point to because they think it shows that the creation days described in Genesis 1 are taken literally in other parts of scripture, so we should take them literally too. The verse from Jude is a favorite of apologetics groups as it gives them permission to "contend," or fight, for the truth of the faith. And my professor's admonition to defend the faith "non-fideistically" meant to produce good reasons for our faith ("fideism" is the position according to which you can just believe without having evidence). So, taken together as an inscription inside the cover of *The Genesis Flood*, the message to me was clear: "This book is what you need to fight the evolutionists and show they are wrong."

Maybe I'm just not a fighter. At least, the kind of attitude this fighting seemed to require was off-putting to me. The book claims, with no room for dissent, "When confronted with the consistent Biblical testimony to a universal Flood, the believer must certainly accept it as unquestionably true." That leads to the very clear choice Whitcomb and Morris believe we all must make: "The decision must be faced: either the Biblical record of the Flood is false and must be rejected or else the system of historical geology which has seemed to discredit it is wrong and must be changed."[1] In other words, you have a choice to make: the Bible or modern science. No question about it. And once you've chosen the Bible—as all good Christians will—you must fight for it.

Fighting for good against evil cultural forces makes for good drama. We love an underdog story in which truth is spoken to power. In the context of evolution and creationism, that has often taken the form of Christian films in which a plucky religious student works up the courage to confront her atheist teacher. The teacher has claimed that evolution shows there is no God, but then the student rattles off a list of supposed problems with evolutionary theory, which evidently the teacher had never considered and doesn't know how to answer. Even if the teacher isn't converted to Christianity on the spot, he has a new admiration for the intellectual rigor of the student.

I recently saw one of these stories in a promo video for a series of classroom resources.[2] The biology teacher challenges the class, saying he's aware some of the students come from families that don't believe in evolution for religious reasons. But he's going to teach students to think for themselves, he says, and asks if there are any questions. The heroine of this short film at first keeps her mouth shut but is eventually goaded by the teacher to speak her mind. She starts in with her questions, which increasingly take the form of an inspiring speech as the background music swells. She rattles off questions about life coming from nonlife, the loss of information from genetic mutations, and the fossil record.

"If evolution were true," she claims at one point, "we'd have millions of 'in-between' creatures running around everywhere, right? And all of the 'in-between' fossils could fit in the back of my Prius."

The claim that there should be "in-between creatures" might seem reasonable on the surface, but it is a misunderstanding of evolutionary theory. The student seems to think that for any two creatures existing today, evolution predicts that there should be an intermediate creature halfway between them. That's the fallacy TV star Kirk Cameron and anti-evolution crusader Ray Comfort fell into when they notoriously teamed up to claim they had easily debunked evolution. Cameron held up a picture of a "crocoduck" and asked why, if evolution is true, we haven't found any of these half-crocodile, half-duck creatures.

As a teenager, I would hear such claims and wonder why they didn't bring the scientific field of evolution crashing down. It must be, I reasoned silently to myself, that the objections weren't as easy and obvious as the crusaders made them seem. I've learned since then that the prediction of evolutionary theory is that for any two organisms today, if you go back in time far enough, there would be common ancestors from which they both descended. For crocodiles and ducks, that's about 245 million years ago,[3] and those common ancestors were not a mash-up of crocodile and duck parts. In other words, they wouldn't have looked at all like the picture Kirk Cameron waved around.

The student in the video is not quite perpetuating the crocoduck fallacy but is instead making a straightforward empirical claim:

scientists haven't found enough "in-between" fossils to fill up a medium-sized hatchback. The charitable interpretation of her claim is that evolution predicts there should be lots of transitional fossils, and we haven't actually found many of these. That was a popular argument by creationists back in the 1980s when their movement was picking up steam. It was wrong then, and with every new fossil find it keeps getting more wrong. But there is a kernel of truth in the student's claim, namely, that fossilization is a very rare event.

The vast, vast majority of living things on this planet leave no trace of their existence other than the DNA they pass on to their offspring. But the wildly incorrect claim the student makes is that scientists have found hardly any fossils that show clear evidence of a transition between two forms.

Consider the Valley of the Whales in the Egyptian desert. It has preserved the remains of more than 1,500 skeletons of creatures that are "in between" the land-dwelling mammals of 100 million years ago and today's whales. These creatures lived some 40 million years ago when there was a shallow sea there. Some of the species still have little back legs that were in the process of becoming obsolete. And not one of these skeletons would fit in the back of a Prius!

Or go to the Hall of Human Origins at the Smithsonian Museum of Natural History in Washington, DC. There is a wall of skulls that have been found from creatures somewhere "in between" our species today and our last common ancestor with chimpanzees. A plaque there says that fossil remains of more than five thousand individuals have been found that can be assigned to this in-between period. That number has grown considerably since the plaque was installed, and again, you're not going to fit all of these in the Prius. The skulls in the display are clearly not our own species, and they are clearly not from chimpanzees. They are somewhere "in between."

If the facts are so decidedly on the side of evolution, why does the creationist position persist so strongly in some communities and seem so obviously reasonable to them? The 2015 Pew survey I mentioned in chapter 1 asked a few other questions about science and broke out the results according to the answers given by white evangelicals. We've already seen that when asked whether humans evolved over time or have persisted in their present form from the beginning of time, 65 percent of American adults chose "evolved." When we limit that question to white evangelicals, though, acceptance of evolution drops to 36 percent. That's not too surprising, but the responses to the next question Pew asked ought to make us scratch our heads:

From what you've heard or read, do scientists generally agree that humans evolved over time, or do they not generally agree about this?

Less than half (46 percent) of America's white evangelicals think scientists generally agree that humans have evolved over time. That is pretty remarkable and might be the key to understanding this community. Two things need to be carefully noted about this finding.

First, the question does not ask people what they themselves think about evolution; it asks what they think *scientists* believe about evolution. The problem here is that the views of more than half of white evangelicals diverge widely from reality. When Pew asked the scientists themselves whether humans evolved over time or have persisted in their present form, 98 percent said humans have evolved. And when they narrowed the respondents to working scientists with a PhD in biology or medicine, 99 percent said humans have evolved.[4] That is about as unanimous as you can get in the real world.

Second, I'm not using the data that 99 percent of experts believe evolution happened to argue that evolution must be true. Truth isn't decided by a popular vote or even by a vote of experts. Ninety-nine percent of experts have been wrong before. Rather, I'm using this data to show that more than half of white evangelicals believe there is a massive controversy in the field of evolutionary studies. In the real world of scientific research, the experts argue about some of the details of evolution, and new findings fill in gaps about what they know. But among the people who know the field the best, there just isn't any controversy about whether humans evolved.

The problem here is not that white evangelicals get their information about science from films or other popular sources rather than directly from the scientific research. Almost all of us do that. A very small percentage of people do the actual scientific work of experiments and interpreting the evidence. The rest of us have to find sources we trust. But for those of us who accept the evidence of evolution based on books and articles, or even programs on the Discovery Channel, behind those sources is a huge network of researchers and institutions like universities and museums. These sources aren't trusted by creationists who have been conditioned to believe that all the important answers are in the Bible, in fact in the first book of the Bible: Genesis.

FINDING ANSWERS

The organization Answers in Genesis (AiG) was founded in America in 1994 by an Australian, Ken Ham, to provide "answers to questions surrounding the book of Genesis, as it is . . . the most-attacked book of the Bible."[1] AiG produces resources in print and online defending young Earth creationism. The organization employs hundreds of people and has brought in hundreds of millions of dollars from its two main attractions in northern Kentucky, which are within a day's drive for almost two-thirds of the country's population.[2] That includes me, so I figured I should have a look because I wanted to understand why the people at AiG believe the things they do.

The most recent attraction built by Answers in Genesis is a theme park called the Ark Encounter. The centerpiece of the park is a life-size replica of Noah's Ark, as described in the book of Genesis. In 2016, soon after the park opened, I had the opportunity to visit with a couple of my kids and a biologist friend.

My first impression of the ark was that it is really big. You catch a glimpse of it from the parking lot, but from that vantage point you don't really have a sense of its scale. Then you board a shuttle bus that takes you back over the hills closer to where it sits. From there you can see the whole thing; the people next to it look like little specks, and you can't help but be amazed by its magnitude.

Genesis 6:15 says that God told Noah to build an ark and make it 300 cubits long, 50 cubits wide, and 30 cubits high. A cubit was a unit of measurement defined by the length from your elbow to the tip of your middle finger. The cubit has the obvious drawback for engineers today of varying somewhat from person to person. AiG took it to be on the large end of the possible range (a little over 20 inches), making their ark 510 feet long. It is really impressive in that sense.

But if you take the details of the story literally, it is only natural to ask how Noah—a man the Bible says was six hundred years old!—could have built this massive structure using only the technologies available in the Stone Age.

AiG does not make the obviously false claim that everything

in the Bible should be taken literally. When Jesus says, "I am the true vine" (John 15:1), no one interprets that to mean that Jesus claimed he was a plant! But AiG thinks the "plain reading" of scripture is usually pretty obvious and easy to discern. So when the Bible says the Lord made the heavens and Earth in six days, it means what any normal speaker of English would understand from that. When the Bible says Noah was six hundred years old and built a boat that was over five hundred feet long, that's what it means. And when the Bible says God told Noah, "You are to bring into the ark two of all living creatures" (Gen. 6:19, NIV), that's what it means . . . kind of. The plain reading of "all living creatures" could include plants, fish, and marine mammals like whales. Obviously, though, Noah didn't have to bring whales on board, right? And the plants would eventually all grow back. So the plain reading of "all living creatures" isn't quite right. But it gets trickier.

Even if we restrict ourselves to animals, scientists have identified more than a million different species, and they estimate there are more than 7 million today.[3] The ark, as AiG built it, has just over 2 million cubic feet of space. This means there would be just one cubic foot of space for seven animals if the ark housed two of each species. That might work okay for the worms and beetles but not so well for elephants and dinosaurs (which AiG claims lived alongside humans in the days of Noah). So the plain reading of scripture has to be discarded here for a much different interpretation. Instead of the millions of different species we see today, AiG claims that what God really meant was for Noah to bring fourteen hundred "kinds" of animals onto the ark.

In their defense, the human authors of scripture didn't talk about "species," which is a modern scientific concept. They just talked about different kinds of animals, which is a commonsense view of what they saw around them. But to reduce the number of animals to those that could reasonably fit on their ark, AiG has to take "kind" as a much higher taxonomic level than species. For example, the ark didn't have to have a pair of Great Danes and a pair of cocker spaniels and a pair of Chihuahuas and all the rest; there was only one kind of dog. The same goes for cats—though not just house cats but also big game cats are all of the same kind, so Noah only needed a pair of these.

Then after the flood receded, the surviving animals very rapidly evolved into the millions of species we see today. My biologist friend who came along, Joel Duff, has written quite a bit showing that this kind of "hyper-evolution" after the flood doesn't work.[4] No matter how you define "kind," you just can't get the millions of species we see today from only 1,400 kinds of animals leaving the ark 4,350 years ago.

Someone might respond to this by saying, "God could have miraculously made that happen. What's wrong with that?" This is a curious question, and one that might provide a key to understanding how this community thinks, so let's pursue it.

AiG accepts the reality of miracles, so why don't they just say, "We know this story couldn't have really happened through natural causes the way it's described in the Bible, but God stepped in and miraculously enabled a six-hundred-year-old man to make a huge boat with fourteen million animals on board"? Maybe they could say that the ark was bigger on the inside than it is on the outside—think of the tents at the Quidditch World Cup in *Harry Potter*, or the Kingdom of God shed in Narnia's *Last Battle*. Wouldn't that kind of explanation make the story even more of a testimony to God's power?

Or think about the amount of food required to feed all those animals on the ark for a year . . . and then about how Noah might have coped with all the waste on board. This smelly problem was answered in an exhibit with an animated video showing an elephant walking on a treadmill to power a big scooping system that emptied waste into a massive septic tank. More than once, I overheard visitors say something like, "That's pretty cool! I can see how that would work." I had a different reaction: I stood there watching the video and asked my friend, "Why wouldn't God just have the animals hold it in for a year?" Perhaps that sounds like I'm making fun of AiG's proposed solution. I'm not. I mean this seriously as a question that gets to the jarring combination of creative scientific and technical solutions on the one hand, and the wholesale rejection of modern scientific findings on the other hand.

I think their logic works something like this: We want to convince people that the Bible can be trusted. When people today pick up a Bible and read a story like Noah's, they might be skeptical that it really could have happened just like the text says. That kind of doubt will eat away at the foundations of scriptural truth and make people doubt other aspects of what the Bible claims. To really counter these doubts—to contend with them, as the verse in Jude urges—we must show that these stories could have happened just as described. That will convince people that the Bible is telling the truth.

I've never seen the creationist community respond to this question about why they don't invoke miracles instead of developing an alternative science. Maybe their alternative science has become an identity marker that clearly distinguishes them from the rest of the world.

I don't think they recognize, though, some of the consequences of these explanations that strain credulity. When coupled with the tone of certainty and urgency, they have a way of creating a rigid and fragile belief system: things must be this way, and everything depends on it. That doesn't invite curiosity; it stamps it out. I'm afraid that is one of the hallmarks of the evangelical world I grew up in. And yet, over time, I have come to believe in a God who welcomes curiosity, as we'll see later.

As further evidence of squelching questions, consider AiG's other main attraction, which a former colleague and I visited a few years ago. They call it the Creation Museum. It too is more like a theme park than a museum, but that's not quite so obvious. You see the kinds of exhibits found at regular museums that give some detail of what their scientists are working on. We even went to a lecture by one of the staff members that included lots of "scientific" detail. But the conclusions being given were so far outside of the scientific consensus that it felt like an alternate reality.

I'm not claiming that AiG is intentionally trying to deceive people. I'm sure most of the people associated with the organization genuinely desire to do what is right and uphold the truth. But their strategy has not been to persuade the experts but to appeal to the masses.[5] As such, they are not involved in the substance of the scientific process, but only the form.

They give scientific-sounding lectures, they have a peer-reviewed journal (with very carefully selected peers, of course), and they have a museum. What allows their very different "scientific" conclusions to emerge from these, and what makes those alternative conclusions so attractive to the evangelical masses, is the claim that the Bible is the proper starting point for scientific inquiry.

The central claim of the lecture we attended and of the museum in general is that any difference of interpretation of the facts from mainstream science can be explained by your starting point: man's word versus God's word. If you start by taking the word of God as authoritative (instead of the word of the godless scientists who are peddling evolution), then you see how every piece of evidence fits into the picture they present: the Earth is only six thousand years old, dinosaurs coexisted with humans, and a massive flood covered the entire planet and killed everything except Noah's family and the animals on the ark. You get the feeling that all the answers really are in Genesis.

My colleague wrote up an account of our visit for the BioLogos website and included this observation:

> As Jim and I made the long drive back to northern Indiana, we agreed that the most striking feature of the Museum is its insistence on answering everything. Every possible question or mystery is defeated by a clear, simple presentation of the Bible's message. Over and over, the Museum is insistent that the worldview presented by Answers in Genesis can answer all of life's questions with different combinations of the same short, snappy, unassailable one-liners.[6]

Answers in Genesis is currently the most visible of the creation science groups, but there are others. They all start by taking their "plain reading" of scripture as definitive and then force science to conform to that interpretation, allowing no room for doubt or dissent. Remember the line from *The Genesis Flood*: "The believer must certainly accept it as unquestionably true." For lots of people, though, the young Earth explanations and interpretations have become too flimsy and detached from reality. By the logic of that claim, if you don't "certainly" accept these explanations as "unquestionably" true, that must mean you're not really a "believer." The unintentional effect of definitively answering every question—of contending for a particular view as though everything depends on it—is to drive people away.

MY CHURCH LIED TO ME

My church lied to me." That's how Christian author Philip
Yancey answered on the podcast when I asked whether he
was frustrated with how his church conditioned him to think
about science. Yancey has written a lot of books critiquing the easy
answers given by the Christian community to difficult questions.
He grew up in a fundamentalist church in Atlanta that denied
there were ever dinosaurs, and they preached that Black people

were cursed and could never be leaders. Yet when Yancey won a summer fellowship at the Centers for Disease Control, he discovered his mentor would be a Black man with a PhD in chemistry. That's when he realized his church had lied about race. "Lied" is a pretty strong charge, but they were obviously wrong. He went on:

> And if they're wrong about race, maybe they're wrong about evolution, maybe they're wrong about the Bible, maybe they're wrong about Jesus. And it was a huge crisis of faith. And the only way I dealt with it was just to back off from faith for a period of time. It took me—and is taking me—a long period of time to sort out. That's what I do as a writer, to sort out what is worth keeping and what needs to be shed because the church doesn't always get it right. And if you get it wrong on science, if you get it wrong on one of these topics, you are opening the door to people just dispensing with everything that the church taught.[1]

That's a pretty good description of the way religious organizations push people away when they give unbelievable answers to sincere questions.

There is only a slight difference between answering a question with a far-fetched answer that must certainly be accepted as unquestionably true, and not answering the question at all. Toward the end of my time as a professor, a student came into my office in tears. She had just come from a Bible class where she asked a question about the way our faith tradition typically interpreted some passage. I don't even remember now which passage—probably one about what women are allowed to do in the church, or why God seemed so mean to everyone but Israel in the Old Testament, or how the Bible's creation story doesn't seem to fit with what science has discovered. But I do remember the answer she said the professor gave: "You shouldn't be asking questions like that."

I wish what that student experienced from a Bible professor and what Yancey experienced from his church were isolated experiences. But I'm afraid in my community of origin, they are not. There seems to be a fear that anything other than putting a swift and definitive end to questions would only cultivate doubt. To show how radically wrong and counterproductive that approach is, we need to look at a few more statistics.

One of the most dramatic cultural shifts in the United States during the twenty-first century has been the rise of the "Nones." These are people who report being affiliated with no religion. Pew started asking a question about religious identity in 2007, finding then that 78 percent of Americans identified as Christian, 16 percent were Nones, and 5 percent selected "Other Religions."[2] They have asked the same question regularly since then, and there is a pretty remarkable trend. By 2021, the percentage of Christians was down to 63 percent, while the Nones went up to 29 percent ("Other Religions" stayed about the same).

Identifying as a None is not the same as not believing in God. Some Nones do believe in God, just not from within the structure of an organized religion. Such people might adopt the label "spiritual but not religious," which has become popular. But the number of US adults who report not believing in God is also on the rise. Professed belief in God was well above 95 percent throughout the second half of the twentieth century. By 2011 it had dipped to 92 percent, and in 2022 it hit a new low of 81 percent.[3]

This data forces us to ask why people are leaving religious faith or questioning their beliefs about God and a spiritual realm. The reasons are undoubtedly complex. But drawing from several different surveys, we can at least point in the direction of a plausible answer.

A study conducted by the Fuller Youth Institute from 2004 to 2010 found that about half of all students who grow up in the church walk away from religion after they graduate from high school. Through in-depth interviews with both those who left religion and those who stayed, they found that one of the leading indicators for those who abandon their faith is that they didn't feel they had a safe place to ask questions. On the whole, those who did feel free to ask questions and raise doubts showed greater faith maturity after high school.[4]

So what kinds of questions do young adults have? In 2019 the Barna Group asked eighteen- to thirty-five-year-olds, "What, if anything, makes you doubt things of a spiritual dimension?" Among the answers they gave were "hypocrisy of religious people" (31 percent), "human suffering" (28 percent), "unanswered prayer" (18 percent), and "education" (16 percent). But most interesting to me, science was also cited as a cause of doubt—in fact, it was the second highest on the list (30 percent), barely behind hypocrisy.

That wasn't too surprising to me because I have talked to lots of college students over the years, and the topic of science comes up very frequently. What did surprise me was how pastors answered a similar question. In 2021 BioLogos commissioned Barna to conduct a survey of US pastors. We listed the reasons the eighteen- to thirty-five-year-olds gave for doubting spiritual things and asked:

All of the following statements are topics that make young adults doubt their religious faith. Which three do you believe are the most common reasons that young adults who grew up Christian leave their faith?

When the results were tabulated, the pastors also picked "hypocrisy of religious people" as the leading reason. But science ranked only seventh on their list (and eleventh on the list for mainline pastors). They thought things like friends, their upbringing, and past experiences with a religious institution were far more doubt-inducing than science.

To summarize, one of the big reasons young people walk away from their religious faith is that their questions are not taken seriously. They have questions specifically about science, but church leaders don't seem to realize this. Organizations that do realize it, like Answers in Genesis, give answers that seem bizarre and outlandish compared to mainstream and well-confirmed scientific views. The result is that people grow up in religious communities like these with a particular view of science that is so tightly wedded to a particular view of the Bible that it is essentially a package deal. Then, when they get out into the real world (or even just watch a nature documentary) and realize that their view of science is clearly wrong, they throw out the whole package. They dispense "with everything that the church taught," in the words of Philip Yancey.

Hearing from lots of former students who went down that path made me want to start addressing these topics with my current students, giving them a place to ask questions and showing that science doesn't have to lead them away from faith. I wanted my students to see that they didn't have to give up their faith because of science. As I wrote in the introduction, proclaiming that message is what led to my departure from the college.

Before my interrogation in the college boardroom, faculty members and the college board's investigatory committee held a series of meetings on evolution over many months. For one of those meetings, I had been asked by the president of the college to prepare some readings on the range of views Christians have taken on the science of origins. I assigned a series of articles from the BioLogos website called "Science and the Bible."[5] The articles do not argue for the truth of one position but simply sketch the history of how different views developed. Learning about this history should help any reader gain an appreciation for the range of views that Christians have found plausible, and it should undermine any certainty that their own view is the only legitimate Christian position. Or so I thought.

For the meeting itself, it was suggested that it might be better if I did not attend—what to do with me was going to be the main topic of conversation. That was frustrating, but I thought some of my faculty colleagues could represent and defend my views capably. However, after the meeting, one of those colleagues came to me very discouraged about the conversation. One of the committee members started making a speech that was clearly uninformed about the range of Christian views. Another of the committee members asked him, "Did you read those articles Jim sent us?" He replied, "I don't need to read that stuff; I've read the Bible."

So that was the level of "conversation" some on the committee were willing to have. These were the men (yes, they were all male) who were setting the policy that faculty members must adhere to—and who were shaping the curriculum for all our students. And they believed the Bible was sufficient to guide them, even in matters related to science.

These men wouldn't look to the Bible for answers about how to fix a carburetor or how to understand the periodic table of elements. Clearly, they acknowledge limits on what kinds of questions the Bible answers for us. But they would say that the kinds of questions the Bible doesn't address are not important for our understanding of salvation. But questions about the origin of humanity are different.

I once interviewed a prominent Christian apologist about his views on the necessity of holding to a literal Adam and Eve. His claim was that Jesus refers to them, and we can't think that Jesus would be wrong or misleading. When we were done recording, I pushed him further on this, saying that Jesus only says, "the one who made them at the beginning made them male and female" (Matt. 19:4), which is a very tenuous connection to claiming all humans must have descended from just two people who themselves had no ancestors; and furthermore, Jesus is recorded to have said things much more clearly that are in fact wrong, like the mustard seed being the smallest seed on Earth (Mark 4:31). And he appears to have endorsed a scientific theory of vision that was current in his day but that we now know to be false: he said, "The eye is the lamp of the body" (Matt. 6:22), which refers to the theory that our eyes see by emitting light. What should we think about these claims by Jesus?

The response of the apologist was that Jesus was simply accommodating the beliefs of his audience by referring to these things.

"Okay, but then why can't we say the same about 'made them male and female'?" I asked. He responded, "Because that has to do with salvation history, and so we must take what he says about it as true for all people and cultures."

That's a pretty sophisticated interpretation, and I'm not saying it can't be consistent. But that sort of gerrymandering with what the Bible says caused me a good deal of discomfort. The plain reading of passages wasn't always so obvious.

So what was I supposed to do with evolution and the Bible? When I was confronted with having to choose between my career and what I believed was true, I saw I needed some help—maybe even a mentor—if I was going to overcome the challenge of how to understand the Bible alongside the science of evolution. I'd find that mentor in C. S. Lewis.

5

FINDING DEEPER FAITH
WITH C. S. LEWIS

After I finished my education degree, my wife and I taught in a mission school in Sierra Leone, West Africa. That was an exciting place to be, but after the sun went down there wasn't much to do but read books by lantern light. The school had a well-stocked library, and I made it most of the way through the shelf of nineteenth-century literature. I liked the big ideas from Tolstoy, Melville, and Dostoyevsky. Somehow in the conjunction between these stories and the analytic processes of math and science from my undergraduate education, out popped a desire to study philosophy.

I actually wrote a letter to the well-known apologist I mentioned at the end of the previous chapter, asking him what he'd recommend for someone like me who wants to study philosophy. He responded that I should get the nine-volume history of philosophy by Frederick Copleston, read it, and outline it. When we came back to the States, I did just that and then went off to graduate school, specializing in the philosophy of science.

That was when I began looking into the evidence for evolution. Eventually, I made it to genetics, which was particularly eye-opening; the interrelatedness of all species through their DNA seemed beautiful and elegant. I didn't have much difficulty in affirming that evolution is the best scientific explanation for how life on Earth developed. But I wasn't sure how to square that with what I believed the Bible says about God as the creator. Was there a way to affirm both the science of evolution and the theology of the Bible? I didn't know how it could be done.

At some point during graduate school, I returned to my undergraduate institution and spent some time with the professor who gave me a copy of *The Genesis Flood*. I shared with him that I was rethinking the relationship between science and the Bible. I needed a more sophisticated way of understanding the authority and inspiration of scripture if I was going to continue believing it's not just another ancient book. I needed someone who could help guide me into seeing the contemporary relevance of that ancient text. My professor did not become that person.

Instead of helping me see options for how evolution and the Bible might be reconciled, he doubled down on their incompatibility and urged me to choose. He meant well, of that I am convinced. He wanted to show me how I could keep my Christian faith (I wanted to help my students the same way once I became a professor). But he believed that evolution is utterly incompatible with Christian faith. Once you are convinced of that, there is no other option than to show that evolution must be wrong.

In the months after our meeting, he sent me long missives about the obvious ridiculousness of evolution. I still have this correspondence and pulled it out recently to see how it strikes me today. There is no doubt he had genuine concern and even love for me, but his urgency and desperation come through too. He provided lists and lists of quotations from people (most of whom I have never heard of) supposedly debunking evolution. One set of these included ninety quotations (I can't help noting that all of them were from males), including the following representative sample:

"Evolution is one-tenth bad science and nine-tenths bad philosophy." *George Frederick Wright,* The Other Side of Evolution: An Examination of Its Evidences

"Evolution has severe problems with scientific evidence. It goes against the evidence. It goes against a direct reading of scripture, has logical problems, and is theologically weak." *Donald Chittick,* The Controversy

"I believe that one day the Darwinian myth will be ranked the greatest deceit in the history of science." *Søren Løvtrup,* Darwinism: The Refutation of a Myth

These statements were supposedly about the science of evolution. I suspect that most of them, though, came from a fear that evolution would undermine the authority of the Bible. Here is another one:

"The biggest problem for those who are theistic evolutionists and who destroy the historicity of the early parts of Genesis has to do with the role of the Holy Spirit as the divine author of scripture. Since the Holy Spirit is omniscient, why in the world would He tell us what He knows is not true and why would He hide from us the truth of how things actually happened?" Harold Lindsell, The New Paganism

I shared that fear at the time. But the science I was learning didn't bear much resemblance to the way these anti-evolution crusaders described it. It really did seem compelling to me. But if the Bible is the Word of God, then why wouldn't God just tell us there, "I created things through the process of evolution"? I didn't know any way of understanding scripture as authoritative and inspired by God that would allow me to accept the science of evolution.

C. S. Lewis was a British literature professor who has had a significant impact on American Christianity. For some reason, when you ask about C. S. Lewis in Britain, people say, "Oh, that chap who wrote the Narnia books for children?" and they don't know much else about him. Over on this side of the pond, we love Narnia, but Christians have also found his nonfiction writing fascinating. *Mere Christianity* continues to sell well, and lots of American Christians point to reading it as an aha moment. Through it they came to understand their faith in that "non-fideistic" sense my college professor admonished me to follow.

At some point during college, I read *Mere Christianity* and found it interesting. But I didn't really have the aha moment so many others had. It was answering questions I wasn't really struggling with. The Christian faith already seemed rational to me—so much so that I thought anyone who didn't accept it must be suffering from some sort of cognitive malady.

It wasn't until graduate school that I found myself surrounded by really smart people who didn't believe the same way I did. The Bible didn't function as a giant answer book for them; in fact, they helped me see that when you treat it that way, you get some obviously wrong answers. But I was not convinced that the Bible was just another ancient book put together by ignorant people. It was more than that to me—I continued to believe that the Bible was inspired by God. I just didn't understand what that meant. And I feared that if I dug into it, I'd find that I couldn't rationally continue to believe it.

What I finally found, though, was Lewis's short book, *Reflection on the Psalms*. That was my aha moment. In that book, Lewis wonders how to understand the inspiration of scripture when so many of the Psalms seem to be advocating for things we don't think are acceptable behavior today. Did God really inspire the passage where the psalmist asks that God dash his enemies' babies against rocks (Ps. 137)? That seems like the kind of thing you might write in a fit of anger, not something that divine inspiration from a loving God would bring about.

We could ask the same thing of passages in other books of the Bible. Did God inspire the passage that instructs parents to take rebellious sons outside the city gate and have them stoned to death (Deut. 21:18–21)? What about 1 Corinthians 14, where the Apostle Paul says that women should be silent in church? Or 1 Peter 2, which says that slaves should obey their masters even when they're dishonest? What about all those passages that instruct us to greet each other with a kiss? Did these instructions really come from God?

I remember as a kid coming across these kinds of passages and asking why we don't obey them. Why does "The Bible says it, and that settles it" only apply to some passages and not others? The plain reading of scripture that my community advocated cannot be applied evenly, and the answer I usually got was something about it being different times back then: those passages were addressing cultural issues that we no longer have today. Then I'd ask, "Okay, then how do you know which passages apply to us today and which ones don't?" I wasn't always the Sunday school teacher's favorite student.

My community thought that to take scripture as authoritative, and not as just another ancient text, we had to believe the very words of scripture were given by God. It's not often said so explicitly, but I grew up with the feeling that God had whispered the exact words of the Bible into the ears of the human authors who wrote them down. Maybe this wasn't an audible dictation, but inspiration certainly meant God was impressing these very words onto the authors. Lewis calls this a "top-down" understanding of the inspiration of scripture.

When you think the Bible is a top-down text, problematic passages are hard to deal with. But Lewis didn't think God simply whispered the words of scripture to human stenographers. Instead, he writes:

> *The human qualities of the raw materials show through. Naivety, error, contradiction, even (as in the cursing Psalms) wickedness are not removed. The total result is not "the Word of God" in the sense that every passage, in itself, gives impeccable science or history. It carries the Word of God.*[1]

So how did these texts come to be? The same way that other texts of the times were written. People put pen to paper (or quill to parchment) and wrote. That doesn't mean God didn't guide the authors in some sense. But the important point for those texts becoming "sacred" was not the method of their production. Rather, it was what happened to them after they were produced. They were "taken up" into the service of the divine. Lewis calls that a "bottom-up" understanding of the inspiration of scripture, and it makes a lot more sense to me. The human-generated words themselves were not infallible, but they bore witness—imperfectly—to a transcendent and perfect reality. Just as God used imperfect people to accomplish the divine plan, so too God could use imperfect texts to accomplish the divine plan.

This way of looking at the Bible inspired me to learn more about C. S. Lewis. My wife and I even made a pilgrimage to Oxford, England, to visit the places he lived and worked. His home, called the Kilns, is available to tour by appointment. So we booked a slot one afternoon and showed up like pilgrims hoping for some holiness to rub off on us. We walked around the garden and down the paths he used to walk, and we sat at the desk where he wrote out the manuscripts of his books in longhand. Then we went to the cemetery at Holy Trinity Church where he is buried. We stood by the gravestone that commemorates him and his brother.

Our next stop was Magdalen College, Oxford, where Lewis taught for a number of years. And finally, we went into town to have dinner at the Eagle and Child. There we sat where Lewis, J. R. R. Tolkien, and other writers known as the Inklings often met to discuss their work.

I'm not sure what we expected to feel or experience from visiting these sites. Nothing mystical or magical happened. In fact, just the opposite. Charles Foster, in his book on pilgrimages, *The Sacred Journey*, says that pilgrimages are the cure for one of the deadliest heresies in the church: Gnosticism. The Gnostics tried to persuade people that only the spiritual realm is good and holy and that the physical realm of flesh and blood is the source of all evil. But a pilgrimage takes you squarely into the realm of flesh and blood. Sometimes it's your own flesh and blood that come to the fore, when the difficulties of the journey impose themselves (I'll have some stories about that later). But a pilgrimage like ours to Oxford put flesh and blood on our idea of C. S. Lewis. He was a real guy who walked on paths, sat in an uncomfortable chair, drank in a pub, taught squirrelly undergraduates, and died. Standing just six feet over his actual decaying bones makes you realize that he too was a regular mortal.

The view of scripture I got from Lewis also forces us away from any Gnostic tendencies to think about the Bible as a purely spiritual book: it forces us to confront the "flesh and blood" of the Bible. I have found this helpful for sorting out the Bible's relationship with scientific claims today.

In the same book, Lewis considers the apparent problem of similarities between the biblical accounts of creation in Genesis and creation accounts from other cultures of the time. He doesn't see the need to suggest that the Bible's creation story was produced in a different way than those other stories were. Lewis explains:

> *Thus something originally merely natural—the kind of myth that is found among most nations—will have been raised by God above itself, qualified by Him and compelled by Him to serve purposes which of itself it would not have served. Generalising this, I take it that the whole Old Testament consists of the same sort of material as any other*

literature—chronicle (some of it obviously pretty accurate), poems,
moral and political diatribes, romances, and what not; but all taken
into the service of God's word.[2]

So God didn't drop the exact words of the Bible down from
heaven. People in an ancient culture wrote what they experienced
and believed, and their stories were taken up by God to be used
in their religious communities to communicate important truths.
That helped me see biblical texts as emerging out of a specific culture and leads to a different understanding of the Bible: rather than
a universal answer book, it is an ancient library full of wisdom from
the perspective of the ancient cultures in which it was written.

Reading the Bible this way is not as straightforward as "The
Bible says it, and that settles it." It is messier, in a sense, because
there is no asterisk in the text that alerts us to when we should
take something literally and when we shouldn't, or when an in-
struction was limited to the original audience and when it applies
to everyone. We have to do the hard work of interpretation. But
it is more honest and more accurate. It makes more sense of what
we find in the Bible itself, and it drastically reduces the perceived
tension between the Bible and science.

With that understanding of the Bible, I felt like I could keep
going on my journey toward resolving the tension I felt between
science and faith. But there were more challenges to come.

PART II

TIME

6

GOD'S WEEKLY PLANNER

Evolutionary theory needs time to work—a lot of time. Billions and billions of years, in fact. Trying to wrap my mind around that much time created challenges for me on a couple of levels. First, my new understanding of the Bible as a bottom-up sacred text still left me with questions. Even if I didn't look to the Bible to answer scientific questions like how old the universe is, I had been conditioned to be suspicious of science operating independently of the Bible. I was ready to learn some of the actual science by which these claims are made about the incredibly old ages of the Earth and universe.

More significantly, I needed to come to grips with what this vast amount of time meant for God's creative process and priorities and where humans fit into them. The challenge of time is not simply noting a discrepancy between a few verses of scripture and what science has discovered about the world. That now seems to me to be rather easily dealt with. Rather, there is the bigger picture of our identity and our place in the grand scheme of things that seems to emerge from the Bible, and a different picture of ourselves that seems to emerge from the long eons of our universe's history. That's the real challenge of time, and I'll try to sort it out in this part of the book.

Scientists tell us today with a great deal of confidence that our universe is 13.8 billion years old. Our minds are not well suited to comprehending that much time, and we easily lose perspective of what happened when. So let's compress all that history into a more familiar timescale: a week. That was how the author of Genesis 1 made sense of our origins. I'm not suggesting we take that literally. I just want us to get a feel for the relative timing of things with what I'm calling God's weekly planner, and then we'll think about the implications of what we find.

To help orient us to the timescale of God's creation activities, I'll list some noteworthy events in the history of our universe. For each one I give the best scientific estimate of when it actually occurred (using the acronyms *BYA*, *MYA*, and *KYA* for billions, millions, and thousands of years ago); I'll also indicate when it would have occurred if the history of our universe were compressed into a week.

In case you're wondering about the math behind this compression, here are the wonky details:

- The universe is 13.8 billion years old.

- For any event that has occurred, we can determine when it occurred as a percentage of the entire history. For example, if something happened 8 billion years ago, that means it happened 5.8 billion years after the beginning of the universe (13.8 minus 8). Dividing 5.8 by 13.8 gives us 0.42, or 42 percent.

- To convert that to our week's timescale, we have to calculate when 42 percent of a week has passed. In seven days, there are 168 hours, and 42 percent of that is 70.56 hours.

- Finally, we convert the hours to days, hours, and minutes on our weekly calendar and get 2 days, 22 hours, and 34 minutes. So that event would have occurred on Tuesday at 10:34 p.m.

THE CREATION OF THE UNIVERSE (13.8 BYA): SUNDAY 12:00 A.M.
Just as the clock hits midnight, our new week begins. This is the moment of the big bang, when all matter and energy explode into time and space.

THE MILKY WAY FORMS (11 BYA): MONDAY 9:36 A.M.
By the next day, our local galaxy of stars has formed. Individual stars in the galaxy are born and eventually die out. Today there are

more than 100 billion stars "living" in the Milky Way. If you're away from city lights on a clear night, you can see one of the arms of our spiral galaxy. The individual stars that make up the arm are too far away to see with the naked eye, so it just looks kind of "milky."

OUR SUN FORMS (4.57 BYA): THURSDAY 4:22 P.M.

We skip all the way to Thursday afternoon. It seems like not much has happened for billions and billions of years. But there were lots of stars going through their life cycles, which—importantly for us—generated the heavier elements that make life possible. Stars burn by fusing atoms together, a process that gives off lots of energy. And the fusion creates new elements. If two hydrogen atoms (with one proton each) fuse together, they create a helium atom (which has two protons). If these fuse, they make other elements. When the star dies, it spews out all those elements in a supernova, they are collected together again by gravity, and a new star can form. From the chemical composition of our sun, we know that it must be at least a second-generation star or possibly a third-generation star.

EARTH FORMS (4.54 BYA): THURSDAY 4:44 P.M.

Not all of the material spewed out by supernovas becomes new stars. Some of it collects into planets instead. And just about 30 million years after the sun forms (a mere twenty-two minutes on the timescale of a week), there is a big liquid ball that eventually cools—at least on its outside—into a rock that we call Earth.

FIRST LIFE ON EARTH (4.0 BYA): THURSDAY 11:17 P.M.

It is surprisingly difficult to define life, and that makes it difficult to pinpoint just when it first occurred. Microfossils have been found in rocks that are dated somewhere between 3.77 and 4.28 billion years ago. The best models of our early planet suggest it was too hot and inhospitable for life for several hundred million years. But once Earth cools and enough water forms, single-celled life comes very quickly after that. If we put it at 4 billion years ago, that's 71 percent of the way through our week.

MULTICELLULAR LIFE (800 MYA): SATURDAY 2:16 P.M.
Once again, it seems like not much has been occurring for a really long time. The first fossils of more advanced, multicellular organisms don't appear until more than halfway through the final day of our week. Then things start happening more quickly (relatively speaking!).

FIRST TREES (400 MYA): SATURDAY 7:08 P.M.
It's close to sunset on the final day of our creation week when trees appear on Earth.

DINOSAURS (230–65 MYA): SATURDAY 11:12 P.M.–11:23 P.M.
Everyone's favorite ancient megafauna, the dinosaurs, have a remarkably long run—though only eleven minutes on our timescale. Most of the dinosaurs stop appearing in the fossil record just after a massive meteor strikes the planet (65 million years ago) and causes sweeping environmental changes. Some of the smaller dinosaurs evolve into birds.

PRIMATES (55 MYA): SATURDAY 11:23 P.M.
Small mammals exist alongside the dinosaurs, but it takes the extinction of dinosaurs before bigger mammals evolve. (They would

have been easy meals for a *T. rex*!) Some of these are primates, which are noted for having bigger brains, relying more on the sense of sight than smell, having opposable thumbs, and being highly sociable. Today primates include monkeys, lemurs, and apes.

LAST COMMON ANCESTOR WITH CHIMPANZEES AND BONOBOS (6 MYA): SATURDAY 11:56 P.M.

It's getting very late on Saturday night by the time a group of ancient apes separates into two distinct lineages. One lineage eventually evolves into today's chimpanzees and bonobos (they in turn split from each other 1 or 2 million years ago). The other is the ancestral line that eventually becomes us. Chimpanzees did not evolve into humans (as is often misunderstood by celebrities on social media!). Rather, if you go back in our family trees about 300,000 to 400,000 generations ago, you reach a population of common ancestors from which we both descended.

HOMO SAPIENS (300 KYA): SATURDAY 11:59 P.M. AND 47 SECONDS!

There are quite a few other identifiable species between us and our last common ancestor with chimps, and we'll meet some of those in the course of this book. But the species we are today—*Homo sapiens*—doesn't appear until a mere thirteen seconds before the end of the week.

If the remarkableness of this week hasn't struck you yet, take a look at a visual representation of it on a weekly calendar. Just for fun, let's say this is God's weekly planner.

God's Weekly Planner for Creation

Sunday	Monday	Tuesday	Wednes-day	Thursday	Friday	Saturday
12:00 A.M. Create the universe with a Big Bang						
	9:36 A.M. Create the Milky Way Galaxy					
						2:16 P.M. Create multi-cellular life
				4:22 P.M. Create the Sun **4:44 P.M.** Create the Earth		
						7:08 P.M. Create trees
				11:17 P.M. Create the first life on Earth		**11:12–11:23 P.M.** Let dinosaurs roam Earth **11:23 P.M.** Create primates **11:56 P.M.** Separate apes into separate lines **11:59:47 P.M.** Create *Homo sapiens*

Of course, we could add a bunch of other events to this week of creation, but almost all the events that interest us would be packed into the last hour of the week. There doesn't seem to be much going on during the first six days.

It's often noted by time-management gurus that our calendars are a true reflection of our priorities. If that's the case, we have some rethinking to do about what God's priorities are! More on that in chapter 8, but first let's consider how we know these dates are correct.

7

KEEPING TIME

Having been satisfied that the Bible doesn't speak to the scientific dating of our origins, I was still curious about whether scientists were correct in their claims of a very ancient Earth and universe. Scientists have been wrong before, and I still had ringing in my ears the charges I heard growing up about scientists being brainwashed with the secular philosophy of evolutionism. To fully answer the challenge of deep time, I needed to understand the science behind scientific conclusions.

Our ability to think about the past and plan for the future has led to various ways of keeping track of time. Whatever form they take, whether we call them clocks or calendars, these methods depend on regularly occurring natural phenomena that we can observe and measure. Our latest digital clocks are designed around a quartz crystal that oscillates at a very consistent rate (sixty times per second) when electricity is applied to it. The electronics in the watch count those oscillations and convert them to seconds. Even more impressive today is the atomic clock, which can detect the 9,192,631,770 oscillations per second of a cesium atom and keep time very accurately.

But our ancestors didn't need fancy technology to keep track of time; instead, they relied on other easily observable natural phenomena. Watching the sun rise and set allowed them to mark the unit of time known as a day. It doesn't take long, though, before the days blur together in our recounting of them. We need a longer unit of time. The phases of the moon can help to some extent, but lunar cycles don't map very well onto the more important annual rhythms of time, especially in temperate climates where the seasons are important for agriculture. Ancient calendars were constructed to keep track of annual events.

On a recent trip to the United Kingdom, some of my BioLogos colleagues and I went to the Orkney Islands, north of the mainland of Scotland, where there are several well-preserved ancient calendars. We signed up for a guided tour of Maeshowe, which is a five-thousand-year-old cairn—a domed structure made of big slabs of rock and covered with sod. Judging from bones excavated at the site, Maeshowe probably served as a tomb.

It's really quite remarkable that there were people and a civilization on Orkney five thousand years ago. That's centuries before the pyramids were built in Egypt and millennia before the Romans came to the British Isles and built roads for easier transportation. These people had to use boats or rafts of some kind to get to Orkney, transporting animals and timber from the mainland, as the islands didn't have much of either.

There are a number of cairns in Orkney, and Maeshowe is the best preserved. You'd think such a remarkable five-thousand-year-old site like this would be either carefully guarded or highlighted with neon signs. But it would be easy to miss as you drive along the nearby two-lane road. The cairn looks like a grassy mound rising about twenty-five feet above the field. There is a visitor center a mile away where you can get information and board a shuttle bus for the tour.

After a short ride, our shuttle bus parked on the side of the road—there isn't even an official parking lot—and our guide led us along a dirt path through a flimsy gate back to the cairn. Next we got to venture inside the cairn itself. To get into the main chamber, there was a long, low passage to traverse. It wasn't like one of those cave entrances you hear about where you have to slither on your belly with your head turned sideways to squeeze through—if it had been, I wouldn't have been doing it! But we had to crawl on hands and knees for more than thirty feet to get to the central room, where we could stand comfortably.

The orientation of the passageway is clearly intentional. When you are inside and look back out through the passage, you can see a standing stone (think of those at Stonehenge) about half a mile away. On the first day of the winter solstice, if you're inside the cairn and it's a clear evening, you can see the sun setting precisely over that stone. It's remarkable that these ancient people understood the patterns of nature well enough to build such a precise calendar, whatever its mysterious purpose.

There are other time-keeping devices that don't have to be constructed. One of the most familiar natural calendars is tree rings. In temperate climates with marked seasonal differences, there is a period each year when many species of trees grow more quickly followed by a period when their growth is slower. The slower growth leaves behind denser, darker wood, while quicker growth yields lighter wood. That gives the characteristic pattern of rings seen in a cross section of a tree trunk. In the trees where this pattern is known to happen annually, you can count the rings and determine how many years the tree lived. Because the rings vary in width with favorable or unfavorable growing conditions, a distinctive pattern is left behind, kind of like a fingerprint. Similar trees growing in the same region have very similar fingerprints for the same years. By matching up those patterns in trees of overlapping ages, researchers can create an unbroken calendar far back into the past. There is a sequence of oak trees in Ireland that goes back 7,429 years, and a sequence of oaks and pines in central Europe that goes back 12,460 years.

There is a similar phenomenon of annual layers at the bottom of some lakes that freeze over every winter. When the lakes are unfrozen, the water circulates more consistently, which keeps finer particles like silt in motion, while heavier particles, like sand from feeder streams, still sink to the bottom. When the lake is frozen for a long period of time, the circulation slows and even those finer particles sink to the bottom. That creates visible layers on the bottom of the lake called *varves*. Varves can be counted like tree rings and preserve information about the past. These can easily stretch to 20,000 years ago, and one in Japan goes back more than 50,000 years.

Varves can tell us about the local environment at certain times in history by preserving things like volcanic ash. When there is a nearby volcanic eruption, as much as a centimeter of ash can be preserved in a varve. For more distant eruptions of significant magnitude, microscopic traces of ash that landed on the lake surface and then settled to the bottom can be identified. These layers help scientists assign dates to eruptions that have taken place and altered ecosystems.

Even more important for establishing dates of ancient natural events is radiocarbon, a material in the varves that helps calibrate another form of dating. Radiocarbon, also called carbon 14 (C-14), is a radioactive form (or *isotope*) of carbon, which occurs naturally in the atmosphere. The most common form of carbon is carbon 12, which has six protons and six neutrons. Comparatively small amounts of C-14 form when a nitrogen atom, which has seven protons and seven neutrons, interacts with cosmic rays in the upper atmosphere. That interaction converts one of its protons to a neutron and creates C-14 with its six protons and eight neutrons. C-14 can combine with oxygen to form carbon dioxide, which almost all living things take in as part of their life cycles. Plants absorb it during photosynthesis, and animals eat those plants (or other animals that eat those plants).

While an organism is living, the proportion of C-14 to C-12

in its body is the same as in the environment. Once an organism dies, it stops taking in new carbon. Because C-14 is radioactive, it decays over time back into nitrogen, so the ratio of C-14 to C-12 goes down. The time it takes for half of a radioactive sample to decay is called the *half-life*. For C-14, the half-life is about 5,730 years. So if you find the remains of an animal with only half of the C-14 it had when it died, you know that animal died 5,730 years ago. If it only has a quarter of the C-14 it had when it died, you know it died 11,460 years ago. And so on. The trick is, you have to know how much C-14 was in the atmosphere when the organism was living and absorbing it. And that amount can vary somewhat.

This is where varves and tree rings come in very handy because they preserve organic material that can be dated independently. Scientists can go to a specific year in the layers that have been preserved and see how much C-14 is left, and because they know the exact year of that layer and the half-life, they can determine with high accuracy the original amount of C-14 for that year. Then that number can be used to date other organic material in the area. Radiocarbon dating can be used to date once-living things back to 50,000 or 60,000 years ago. For periods before that, there isn't enough C-14 left to measure it accurately (it has been cut in half ten times by then, leaving only 1/1024 of the original amount).

But there are other kinds of radioactive dating for nonliving things that go back much further. Potassium 40 has a half-life of about 1.3 billion years. It is present in molten lava and decays into argon 40. This method of radioactive dating has been used to determine the age of rocks in Greenland to be more than 3.5 billion years old. That testifies to a very old Earth.

On the scale of the universe as a whole, scientists have managed to decipher other natural timekeepers. The speed of light is constant and quite literally lets us see back in time. The nearest star in our galaxy is about four light-years away, which means what we see now is the way it looked four years ago because that's how long it took the light to get to us. Much more distant stars in our galaxy emitted light that has been traveling for thousands of years, and now with powerful telescopes we can observe other galaxies that are millions and even billions of light-years away.

Another natural timekeeper we've come to understand in the last few decades is the cosmic background radiation that is left over from the big bang—a kind of electromagnetic echo from the early expansion of our universe. This was predicted as early as the 1940s but not detected until the 1960s. Now with advanced equipment it can be measured precisely, and it shows the age of the universe to be 13.8 billion years.

As discussed in chapter 2, some Christians believe you can add up the ages given in the Bible and determine that God created the heavens and Earth just 6,000 to 10,000 years ago. I have talked a number of times to creationist Christians about scientific dating methods and the extremely old Earth and universe they reveal. A common response from them is to claim that God could have just created Earth and the universe with the appearance of age. That is to say, they claim that God really created the heavens and Earth just six thousand years ago, but God made it look mature. God didn't create Adam as a newborn baby, they argue, so doesn't it stand to reason that when God created a fully formed tree it would have rings inside? And distant stars would have been created with light already en route to Earth so we could see them!

OK, *maybe* . . . but this reasoning starts to break down pretty quickly. I remember growing up hearing the Bible story about Adam and Eve, and I liked to ask whether they had belly buttons. That might seem like a silly question, but there's a pretty important point riding on it about how we understand God's intentions. Would God create things with clear and compelling evidence of a history that never existed?

Let's say God created a big, mature oak tree in the Garden of Eden with one hundred years of rings in its trunk. But it wouldn't look authentic if the rings were all exactly the same. So maybe rings 52, 53, and 54 were created to look like years of extraordinary growth—the distinctive fingerprint of tree rings I mentioned earlier. But then God wouldn't have created only one-hundred-year-old trees. There would also be some ninety-year-old trees, and to be consistent they would have to have that same fingerprint in rings 42, 43, and 44. And the eighty-year-old trees would have to have it in rings 32, 33, and 34. And so on.

Not only that, but God would have had to put the right amount of carbon-14 in each of these so that they "appeared" to be the correct age to any scientific instruments that might ever investigate them. And God would have had to put fossils in the ground representing creatures that never actually lived but look just the right age to fit with the sequence of life-forms that has been discovered. And God would have had to create just the right cosmic background radiation so it is consistent with the measured age of the universe. And God would have had to not only put a belly button on Adam and Eve but also give them DNA that matches the interconnected history of life that has been discovered.

If a young Earth creationist persists in affirming all this, we could simply say that scientists today are discovering the details of this past that really only existed in the mind of God but in a sense was created and preserved in all these natural phenomena. If that were true, God would be like a fantasy author who created an elaborate world, stipulating all sorts of details that would later be investigated. It's not impossible that God created in this way, but is it the most likely scenario?

By that same line of reasoning, couldn't we say that God actually created the world just six hundred years ago with all the evidence of a much older universe? Or couldn't we say that God created things six years ago, or six minutes ago, complete with our memories of an imagined past before that?! The evidence would be just the same. But that turns out to be a very elaborate ruse on the part of God! It doesn't seem like God would deceive us in that way. That would make us alter a fundamental belief about God. It became much less problematic for me to accept the findings of science—even if they ran counter to the young Earth creationism of my community—rather than believe that God is a deceiver.

To be fair, though, we have to ask whether our realization that the universe has been around more than 13 billion years leads us to change some other things we've thought about God. We need to wrap our minds around these billions of years that have transpired since the universe began and think about what that means about God and what that means about us.

8

GOD'S PRIORITIES

Now that we've seen God's weekly planner for creation in chapter 6 and accepted the science behind it in chapter 7, what conclusions can we draw about God's priorities?

For one thing, God seems to place a pretty high value on watching fusion in the stars produce heavier elements! From Monday morning to Thursday afternoon of the creation week, that's about all that was going on in the universe. Coming back to real time, that accounts for around 6 billion years of the universe's history, or

40 percent of the time God has spent creating things. That seems like a pretty significant chunk of time for the Creator to devote to something so seemingly *in*significant.

We might also conclude that we *Homo sapiens* are but a brief afterthought. For 99.998 percent of the history of creation, there were no *Homo sapiens*. It doesn't seem like we are the central characters in this drama. What are we to make of this?

Before trying to answer that question, let's make the problem harder. Besides considering how much of the time we have existed in the history of the universe, we might also consider other creatures that by some measures seem vastly more important.

A recent study estimates the population of ants on Earth today to be at least 20 quadrillion.[1] That's more than 2.5 million ants for every human on the planet! Does the number of individuals of a particular species say anything about that species' importance to God? Similarly, if we scale down to microscopic organisms like bacteria, we'll find many, many more individual bacteria than humans, but that's just because they're so small, right? Yes, but even by weight, there is over a thousand times as much bacterial biomass as there is human biomass on the planet.[2]

Or consider this anecdote about British scientist J. B. S. Haldane, one of the leading geneticists and evolutionary biologists of the twentieth century. He was asked by some theologians what he could conclude about the Creator based on what he sees in creation. Haldane had no love for religion and supposedly answered rather cheekily, "God must have an inordinate fondness for beetles." About 1.5 million species of living things on our planet have been scientifically described, and about 400,000 of those are beetles.[3] That's almost a quarter of all living species!

Why would God create so many different species of beetles

compared to our one species of human? We might note that there used to be other species of humans too. In part III, we'll meet some of these other species that have been classified in the genus *Homo* and try to sort out whether they were humans like us. For now, it simply raises another difficult implication of the billions of years of Earth's history and the evolution of life: death and extinction. Those creatures, and many more, no longer exist. It is estimated that more than 99 percent of all the species that have ever existed are now extinct.[4] Is death on this scale really part of God's good plan for the world?

This is quite a list of consequences (or side effects) of the way God has created: stellar fusion going on for billions of years, significantly more ants and bacteria than humans, all the different kinds of beetles, and all the species that have gone extinct. What are we supposed to think about this? Taking so much time to create seems to suggest that we were not God's top priority. If we really are God's top priority, then wouldn't there have been a better way to create us? That was a pretty big challenge to my faith-inspired view of the world. How can we respond?

The original science documentary series *Cosmos* was hosted by astrophysicist Carl Sagan in the 1980s. One of the episodes begins with Sagan sitting at high table in the dining hall of a posh British university. Classical music is playing in the background and a butler in formal attire strides in and places a pie in front of him. Sagan looks at the camera and announces in his distinctive, halting voice, "If you wish to make an apple pie from scratch, you must first . . . invent the universe."[5]

If Sagan is right about apple pies, then surely the same point applies to making human beings. We might still say that God's top priority from the beginning was to create human beings, but this long, slow process with all its waste and death was necessary to bring about the raw materials for creating human beings. Even in the biblical imagery, humans weren't created out of nothing. They were molded from the dust. But to do that you have to have dust, which is composed of those heavy elements that were generated during the long cycles of stellar fusion and supernovas.

In evolutionary terms, if you want to create humans from other primates, you first need primates, and before primates you need other terrestrial tetrapods, and before tetrapods you need fish, and before fish you need single-celled organisms. That process is going to result in a lot of other things besides humans: lots of ants, tons of bacteria, and an incredible diversity of beetles. And there will be a long list of creatures that went extinct along the way.

Why couldn't God just speak the humans into existence, or at least speak the dust into existence? We believe God is all-powerful, and miracles are an important part of the Christian story. Why limit God to working through natural processes when a miracle could have accomplished the same thing without all the time and waste?

It's a good question, one that requires us to think a bit more about the nature of miracles. When we do that, I think we come to a different—and deeper—understanding of what God has been doing for so long in relation to the created world.

For several hundred years now, scientists have been making increasingly detailed observations of the natural world and figuring out how things work. Religious communities like the one I came from tend to feel that these scientific explanations are in competition with explanations that invoke God.

According to the Abrahamic traditions, God created land. But from science we know that land is formed by cooling lava from volcanic eruptions. You can go to the Big Island of Hawaii and see this happening in real time. Scientists have observed and measured the formation of new land in this way all over the planet. Does that mean God didn't really create the land? I don't think so. It just means that there are multiple explanations for the same thing.

A popular illustration of this among science and religion scholars is to ask why a teakettle is boiling. A physicist might explain in terms of a closed electrical circuit, which causes the heating element in the stove to convey heat to the bottom of the kettle, which in turn causes the water molecules to move more rapidly, eventually causing the vapor pressure of the water to be higher than the atmospheric pressure, and so the water boils. That's a perfectly legitimate and scientifically complete explanation for why the kettle is boiling.

But we might also explain the same event on a different level: the kettle is boiling because I wanted a cup of tea! We can call that a personal explanation, because instead of appealing to the laws of physics, it appeals to reasons and personal agency. These two explanations don't compete with each other, and in fact we know more about the situation when we consider both instead of just one or the other.

So too with God's creation. Saying that God creates the land is a personal explanation. Appealing to volcanoes and tectonic plates is a scientific explanation. Having a scientific explanation does not write God out of the picture. In this sense, it is legitimate to see God constantly at work in the world, not only in events we typically call miraculous but in regular ones too. By extension, we can still affirm that God created humans (and all of the matter and processes it would take to bring them into being) and at the same time accept a scientific explanation for how we came about.

If we believe the scientific explanation, then does this mean that God could only have brought about humans through a process that took billions of years? That feels kind of like saying that someone who wants a cup of tea must first craft the kettle out of ore, drill a well to get the water, and then go and gather firewood to burn in order to boil the water (let alone planting seeds and waiting for them to grow into plants bearing tea leaves). Wouldn't the God of the universe be able to do the equivalent of popping a cup of water into the microwave for a minute or so?

Maybe. I can't speak for God. But I think we can look at what has in fact happened and then talk about how that fits with what we believe about God.

It makes my inner evangelical nervous to say God *couldn't* have brought things about more quickly. In part V, I'll talk about some things I believe God can't do. But I'm not so sure creating out of nothing is one of those. I'm confronting instead the fact that God didn't create us out of nothing, but rather as part of a long process. My challenge here is to say something about how that might fit with other things we believe about God, starting with miracles.

I'm not questioning the reality of miracles. For Christians who take the Bible seriously, though, we should actually pay attention to how the Bible portrays them.

Miracles in the Bible were not labor-saving devices for God, as though God said, "I need some dust to form humans out of, and I don't want to have to wait around to produce it naturally, so I'll just zap it into being." The biblical understanding of miracles is that they are signs and wonders that testify to the Kingdom of God and how things work there: hungry people are fed, sick or injured bodies are made whole, and Jesus himself resurrects as the "first fruits" of what will happen at the end of time (1 Cor. 15:20). Even Jesus turning water into wine was not a labor-saving device. The point of doing this miracle was not so they didn't have to wait around until more grape juice fermented, but rather to reveal his glory (John 2:11).

I'm skeptical, then, that God would produce signs and wonders when there weren't any people around to appreciate them. What would be the point? To keep God from getting bored during those billions of years of stellar fusion?

If I could, I would probably have the autumn leaves miraculously fall into neat piles so they wouldn't have to be raked, or have stinky diapers change themselves. But that's just me trying to get out of work I don't much enjoy. I wouldn't miraculously skip over a game just to get to the end of it—playing it is the enjoyable part. Maybe God enjoys the work of creating a universe. Maybe God loves doing that.

RETHINKING TIME AND THE BIBLE

seemed to be moving toward a deeper way of understanding God's relationship to time that doesn't challenge the findings of modern science. But I was still curious about how to interpret what the Bible seems to say about the history of humans and the Earth. Young Earth creationists are pretty adamant that the genealogies given in scripture can be used to date the origin of humanity and even the Earth. The genealogy in Luke 3:23–38 seems to give Jesus's ancestors all the way back to Adam, and they assume (the Bible doesn't actually say) that Adam was created on the sixth day of Earth's existence. Other genealogies even give the ages when sons were born, so you can do the math, make some reasonable assumptions, and calculate that the Earth (and universe) are only 6,000 to 10,000 years old (the range is because many young Earth creationists allow that the genealogies could have some gaps or could have skipped generations). Do we just have to say that the authors of those texts got it wrong? That is one option lots of people take nowadays, and it is a legitimate possibility.

Returning to Lewis's view of biblical inspiration from chapter 5, God is perfectly capable of accomplishing the divine plan through flawed human beings. If we could pin down Moses, Isaiah, or the Apostle Paul and say, "Quick, tell us how old the Earth is!" I wouldn't be surprised if they said, "I don't know . . . maybe six thousand years?" But I also suspect they would add, "But so what?! There's nothing very important riding on that, and it certainly wasn't what we were trying to communicate."

In other words, God wasn't inspiring people to write these texts with the intention of smuggling some modern science into them. So we shouldn't try to mine the Bible for answers to scientific questions like "When did the Earth come into existence?" and "How was it formed?" The Bible might reveal what the authors believed about Earth's history, just like it reveals their acceptance of slavery and polygamy. But that doesn't mean people today who take the Bible seriously must also believe those things.

So what is the Bible communicating about time and history? Let's look at the genealogies and the first chapter of Genesis with the help of a couple of Bible scholars.

There's something fishy about the genealogy in Genesis 5. I don't mean simply that the ages ascribed to the patriarchs are unbelievable. Adam lived to be 930 years old?! Methuselah made it to 969! Yes, these are incredible. But I don't think the right way to interpret the Bible is to let science say, "Those ages are impossible, so the Bible is wrong." Instead we might think of a dialogue between science and the Bible in which science says, "Are you sure that's the best way to interpret this passage? Are there any other possibilities?"

If we let science urge us to read this text more closely, we find something else that is fishy. There are thirty numbers given in Genesis 5 that are supposedly about ages, and all of them end with the digits 0, 2, 5, 7, or 9. You might not think that is too remarkable until you realize that it eliminates half of the possible numbers. It is like seeing a list of thirty numbers that are all even. We wouldn't think that was a random distribution of numbers. In fact, the odds of getting all thirty numbers to end with just these digits in a random distribution of ages—the way you would expect them in a list of ages of how old people are when they had a son or died—are about 1 in 100 million. That should make us suspicious of the claim that Genesis 5 is merely giving a historical report. Something else must be going on in that text.

What were the authors trying to convey? The honest answer is that we don't know. I'm afraid the symbolism is lost to us now. We can guess at some of the significance by seeing all thirty of those age numbers as combinations of 60 and 7 (which were both important and even sacred numbers in their culture): Methuselah's age of 187 when Lamech was born equals 60+60+60+7; Adam's age of 130 when Seth was born can be the combination of 60+60 years (120) and 60+60 months (10 years). And so on for all the other numbers given. But still, we don't really know why such combinations might have been important or what they conveyed.

I've talked to Old Testament scholar Richard Middleton a couple of times on the podcast. For one of those episodes, we talked specifically about the genealogies in the Bible and what they might be conveying.[1] Middleton shared a number of interesting insights, including the use of gematria, which is the practice of assigning a numerical value to letters. If you've ever read Chaim Potok's novel *The Chosen*, you've been exposed to the practice of gematria. One of the main characters is the leader of a Jewish Hasidic community in New York City. He regularly uses gematria to the amazement of his congregation, showing how different words must be related because of their numerical values.

The Bible does some of this too. The most famous example is probably from the book of Revelation: "Let anyone with understanding calculate the number of the beast, for it is the number for a person. Its number is 666" (Rev. 13:18). Many Bible scholars believe this was a covert reference to Emperor Nero, whose name written as "Nero Caesar" in Hebrew has the numerical equivalent of letters that add up to 666.

Middleton says this kind of gematria is also driving the genealogy in Matthew 1. It lists three sets of fourteen generations, and the reason is that it is all based on the name *David*. In Hebrew, the name is composed of the three consonants *dalet, vav, dalet* (transliterating to our D-V-D). These are the fourth, sixth, and fourth letters of the Hebrew alphabet (4+6+4 = 14). I suppose that much could be coincidental, but Matthew's genealogy gets even more mathematically suspicious.

The first set of fourteen generations begins with Abraham. When you do the gematria of Abraham, you get 41. And multiplying 14 (David) by 41 (Abraham) gives us 574, which just happens to be the value you get when you add up the gematria of all the names listed in the genealogy from Abraham to David. Well, not quite . . . you have to change the spellings of some of the names to make the math work out. And that's what Matthew does! He gets most of his names from the Old Testament book of Chronicles, but he has to change them slightly to get the math to work. For example, Ram becomes Aram, Boaz changes to Boas, and Obed has to be spelled Jobed. Then the math works out. And it's the same for the other sets of fourteen generations. That makes us think even more that the genealogies are contrived for purposes other than giving the kind of history we expect from genealogies today.

What about Genesis 1, the classic text that seems to say everything was created in a week? If that's not what it's attempting to communicate, then what is really going on? One plausible interpretation has been advanced by John Walton. He is a recently retired Old Testament professor from Wheaton College outside of Chicago. Wheaton is usually named as one of the premier evangelical institutions in the country—it can't be maintained that they don't take the Bible seriously there. But that doesn't mean that they simply look at the "plain reading" of scripture and say that settles it. Walton has done serious work in understanding the ancient Near Eastern culture and context in which these texts were written. Remember, the Bible didn't drop down from heaven!

Walton wrote a short, accessible book called *The Lost World of*

Genesis One. I remember purchasing it at a conference I attended in New Orleans in 2009. I put it in my bag for the plane ride home, and once I started reading it, I couldn't put it down. It makes so much more sense to read Genesis as an ancient document that arose in a specific ancient culture.

Walton believes that God is ultimately responsible for creating the material of the universe, but he doesn't think this is what the creation texts are talking about. For the ancient Near Eastern mindset, to create something was to assign it a function. So when the text says that God created the heavens and the Earth, the claim was not that there was no material—the atoms of hydrogen and so on—and then all of the sudden that material came into being. Genesis 1:1 itself seems to begin with material already there. Modern English translations note this by rendering the original Hebrew this way: "When God began to create the heavens and the earth, the earth was complete chaos, and darkness covered the face of the deep." The text goes on to describe God bringing order to the material world so that it would function properly for humans.

That means when we ask, "How old is the Earth?" and force the Bible to answer, we're going to get irrelevant (or even misleading) information. When Walton was on the podcast, he gave a really interesting analogy of this.[2] Imagine you have tickets to a play and get to the theater late. You finally make it to your seat thirty minutes after the play has started, so you poke the person next to you and ask, "How did the play begin?"

That person turns to you and whispers, "The play was written in 1938. It was a Pulitzer Prize candidate that year and was very popular on stages—"

"No, no," you interrupt. "How did the play begin here?"

"Oh, sorry," he says. "The set was constructed by the Morris

Construction Company, which is very famous for getting sets to fit into this kind of building. And—"

"No! You misunderstand my question."

"Well, you know that you can't have a play without a script and without a set, right?"

"Okay, but that's not what I'm asking. I want to know what I've missed so far."

"The cast was chosen by—"

"Ugh! What has happened since the curtain went up?!"

No metaphor is perfect, but we see here the same kind of disconnect between questions and answers when people look to the Bible for a scientific account of our origins. It's as though they want an account of how the set was constructed, but the Bible just tells the story of what has happened once the curtain went up on the human drama. And consistent with Lewis's bottom-up understanding of how scripture was written, we get just a part of the human drama from that one particular time and culture. We shouldn't think the Bible gives a universal history of all cultures and times, because it was produced with the limitations of those original authors. But remember that Lewis doesn't stop there. The authors' accounts were "taken up" into the service of their religious community and its tradition, and as such they can be understood to speak truth into all people's experience.

It is grossly misleading to reduce a piece of art or literature to a few declarative sentences saying, "This is what it means." The artist or novelist usually reacts to such reductions with, "If that's all I was trying to say, I would have just said that." Instead, the beauty and complexity of art and literature have to be experienced in their entirety. That experience can't be summed up in words without massive reduction in meaning.

So too with the creation accounts from the Bible. We shouldn't try to force them into a few sentences of what they were really trying to say. These stories have been told and read and worked into liturgical performances for centuries. When we participate in these, we respond to what the Bible has to say about creation and origins. Those responses should include a recognition of the creation's dependence upon a Creator and an understanding of the calling placed upon human beings to represent God to the rest of creation.

Science doesn't dictate how we should interpret the Bible—just like the Bible doesn't dictate how we should interpret scientific results. These are different sources of truth, primarily operating at different levels (like the scientific and personal explanations for the teakettle boiling in chapter 8). But people who want a comprehensive and coherent understanding of reality should consider both and allow them to be in dialogue with each other.

For me, even though science didn't force me to adopt a certain interpretation of scripture, it did suggest as part of the dialogue that I might look again at scripture to see whether what I thought it said was really a good interpretation on its own terms. It's time to bring together some of the insights that let me meet the challenge of time.

FINDING DEEPER FAITH WITH ARUNDHATI ROY AND G. K. CHESTERTON

When I travel to a new place, I often try to find a novel that is set there. A story can help give a sense of place better than the facts and figures you might learn from Wikipedia. So when I took a trip to India's southwestern state of Kerala with a friend who grew up there, I asked him for a book recommendation. We were in the small city of Kottayam, and he told me that the obvious choice was Arundahti Roy's novel *The God of Small Things*, which was set in that area.

I hadn't heard of Roy or the book, despite the fact that it won the Booker Prize in 1997. But she is very well known in India as an author and public intellectual.

I wonder if Roy would find it surprising that she helped me find deeper faith in the light of the long history of our universe? I don't know whether she herself is a person of religious faith. But I've read her novel twice now, and two aspects of it have contributed directly to my overcoming the challenge of time.

First, it helped me make sense of how God isn't in a hurry and isn't concerned with time and efficiency. Roy had one of her characters talk about the 4.6-billion-year-old Earth as though it were a forty-six-year-old woman. She was only eleven years old when the first single-celled organisms appeared and was already forty-five by the time the dinosaurs roamed the earth. Human civilization began only two hours ago on the scale of her life. This was an inspiration for my recasting the 13.8-billion-year history of the universe into the seven days of God's weekly planner in chapter 6.

The second insight from Arundhati Roy can be seen in the title of her book: God delights in the small things. Visiting India is often described as an assault on the senses—the colors, the sounds, the smells, and the tastes all seem enhanced. Roy describes these masterfully. But it is not just to add color to the story; these are the important things. The narrator says, "Little events, ordinary things, smashed and reconstituted. Imbued with new meaning. Suddenly they become the bleached bones of a story."[1]

I can't read that line and not think of the hydrogen atoms in a star, "smashed and reconstituted" into helium atoms and eventually the rest of the elements. These become the bones of the story of life in our universe, of our story. Perhaps we shouldn't focus just on the ultimate meaning or purpose, which from our perspective seems to emerge from these small and insignificant parts. Maybe it's not the devil who is in the details, but looking with more focused eyes we might see the God of small things.

These two points need to be developed in greater detail.

As Roy's forty-six-year-old woman and God's weekly planner show, God doesn't seem too concerned about time and efficiency. That makes me think it's not too helpful to think of God as an engineer. If the point was to produce human beings as quickly and efficiently as possible, then the way God created us would not score very high marks from the engineering guild. Perhaps it took science for some of us to realize this, but we also could have seen it in the Bible itself.

God was content to let the children of Israel wander in the desert for forty years before coming into the Promised Land (maybe it would teach them something!). Think of the centuries that passed before the fullness of time was achieved and Jesus came to set things right. And now Christians have been waiting for two thousand years for the promised return of Christ to usher in the eschaton.

Those are not the plans of an engineer, who would have tried to minimize the time needed to achieve the goals, not allowed them to play out over interminably long eons. So if not an engineer, what is God? Of course God doesn't conform to any of our human categories, but I'd suggest that God the Creator is more like a baker or an artist than an engineer.

Most weeks, I bake my own bread. While this was something I

enjoyed and did from time to time in previous years, when COVID hit, I jumped on the bandwagon of sourdough bread baking and never got off. First I had to develop a natural yeast starter. I could have gotten one from someone else who already had one going, or even (gasp!) used commercial yeast to start it. But I wanted to go the natural way and develop my own. That took several weeks of daily feeding before the starter was ready to use. And even now that it is established, there is still quite a lengthy process to producing a fresh loaf of bread.

If I want bread with dinner on Friday night, I have to remember to take my starter (named Francisco) out of the refrigerator early in the day on Thursday. That gives him time to warm up before I feed him more flour and water to get him prepped for feasting on the dough. Then Thursday night I have to mix up the ingredients before bed so Francisco can feast all night long, converting the sugars into carbon dioxide (which makes the dough rise and creates the lovely holes in sourdough bread) and alcohol (which gives it amazing flavor). Depending on the kind of loaf I'm making, there are several more steps throughout the day on Friday before it goes in the oven. Then it comes out steaming and delicious.

If my goal were simply to eat bread, I could use a recipe with a shorter duration or even buy a loaf of bread. But I don't bake bread because I'm interested in saving time. My bread tastes better than store-bought—you can't convince me otherwise!—and I really enjoy the creative process. I've even gone so far as to grind my own flour from wheat berries (my family worries I might next want to start growing my own wheat!). All this takes me deeper into the details and gives me greater satisfaction in the finished product.

The analogy with a creator God isn't perfect. Us dashing into the grocery store for a ready-made loaf of sandwich bread isn't the same as God instantaneously creating ready-made people out of dust.

But my point is that just as a baker finds joy in the process of creating from scratch, so too might God enjoy the process of creating over time. God might even prefer the end product of that process to something instantaneously made. It takes God further into the details. I can even imagine God taking more pride in creating that way.

Maybe you prefer to think of God more like an artist. Consider the artist Makoto Fujimura, who is a leading contemporary painter. He describes his work as "slow art." He doesn't go to the art store and buy paper and paint. He makes his own paper, and he pulverizes minerals into pigments for painting. He paints in the slow tradition of Japanese art where you lay down a hundred layers before you start painting the subject of the piece. I interviewed him for the podcast and asked about this. He responded that you experience a different sense of time during the creation of slow art, such that eternity is packed into each moment:

> You might be laboring for years and years to prepare something and then, all of a sudden, discovery happens, where there's an explosion, an opening, that would only happen if you had spent that time preparing for that moment. It's no longer slow or no longer tedious because it becomes such an accentuated reality of seeing something new.[2]

Understanding time this way goes a long way toward overcoming the challenge of this part of the book.

The second insight from Roy also helps us shift from the perspective of God as an engineer: we humans aren't the only point of creation. God loves the other stuff too.

To an engineer, the incredibly long time from the beginning of the universe to now may look wasteful and inefficient. But that's only if the whole point of creation were for human beings to exist. We can still maintain that human beings were a really important priority (and I'll have more to say about that in the rest of the book) without thinking that everything else in creation is just a means to that end. I think God takes delight in all the other stuff too. With that perspective, we can look at the same history—with all the ants, bacteria, beetles, and extinctions—and instead of wastefulness and inefficiency, we can see the lavishness of God's creation.

The history of science can be read as humankind discovering that our universe is bigger and more extravagant than we had imagined. European scholars in the Middle Ages thought that the Earth was a cozy place in the center of the universe and that human cultures were all clustered together around the Mediterranean. Then explorers found other continents and people on the "bottom" of the Earth. And then scientists discovered that Earth is one of several planets orbiting our sun. Then they discovered that our sun isn't unique, but the same kind of thing as those thousands of other stars they could see in the sky. And then with telescopes they learned that there are hundreds of billions of stars in our galaxy, and then that there are hundreds of billions of galaxies. Now we are finding more and more planets around other stars, and most scientists wouldn't be surprised to learn that these other planets have life on them too. And maybe there are even other universes?!

It's difficult to maintain that all this is only for the sake of us humans. Most of it has almost no connection to us and is beyond the realm of what we'll ever be able to experience. So why did God make it? The answer, I think, is that God delights in all of it for its own sake. That applies equally to the examples I gave earlier. I can see God saying, "Wow, these beetles are really cool! I'm going to let there be 400,000 different species of them." Even the extinctions might be understood along the lines of God's creative extravagance: all those species couldn't have existed at the same time any more than the thousands of planets we've discovered (with millions more surely to come) could exist in the same space. By creating this way, God has allowed many more kinds of things to exist than could have if creation had happened all at once (or over one week).

Okay, so if it's not all about us, and God doesn't seem to be in a hurry, we still might wonder: 6 billion years of nothing but stellar fusion?! An engineer might liken this to watching paint dry. I should quit making engineers the villain of this story ... To be fair, any adult would probably be bored by stellar fusion after a few days or even minutes. But what if we reclaimed a kind of childlike wonder and delight in it, as G. K. Chesterton asserts?

G. K. Chesterton was a British writer in the first few decades of the twentieth century. He wrote fiction and essays, as well as popular books defending the reasonableness of Christianity. One of these latter books, published in 1908 but still widely read today, is titled *Orthodoxy*. In it he attempts to explain how he came to believe in Christianity.

In one section, he reflects on the tendency of scientific explanation to push God out of natural processes. Too often, he thinks, when we discover a law of nature that shows why things happen over and over, we think it is due to impersonal causal forces. It was only back in our prescientific days, when we thought nature was impulsive and capricious, that we assigned gods to be the causes. In typical Chestertonian style, he flips this on its head. The passage is worth quoting at length:

> *Because children have abounding vitality, because they are in spirit fierce and free, therefore they want things repeated and unchanged. They always say, "Do it again"; and the grown-up person does it again until he is nearly dead. For grown-up people are not strong enough to exult in monotony. But perhaps God is strong enough to exult in monotony. It is possible that God says every morning, "Do it again" to the sun; and every evening, "Do it again" to the moon. It may not be automatic necessity that makes all daisies alike; it may be that God makes every daisy separately, but has never got tired of making them. It may be that He has the eternal appetite of infancy; for we have sinned and grown old, and our Father is younger than we. The repetition in Nature may not be a mere recurrence; it may be a theatrical ENCORE.*[3]

Watching sunrises on billions of planets in the galaxy doesn't get old for God. And perhaps God is watching closely enough to see inside those stars where elements are fusing into others: "Do it again, hydrogen! Make another helium atom, another oxygen atom! Then blow them out into space in a supernova, and let them collect again into other stars and planets. Then do it again!" Maybe that's not boring or insignificant from God's perspective, but amazing and endlessly fascinating to watch. Remember the lesson from the boiling teakettle: having a scientific explanation for these things does not mean we can't also understand the personal explanation of God's delight as a true description of the event.

So how do we overcome the challenge of understanding how vast eons of time fit into the bigger picture of creation and our place in it? There is no problem when we realize that God doesn't seem too concerned about time and efficiency. The history of creation doesn't look at all like a Henry Ford assembly line where things are made as quickly and efficiently as possible. It doesn't look like a sprint to the finish line. The history of creation looks more like God has taken a leisurely stroll. It's a journey, and God's not worried at all by stepping off the main road to explore some side alleys, to go out and watch the sun set over and over and over again. God delights not just in stopping to smell the roses, but in stopping where there are no roses yet and watching them grow.

But what about our seemingly minute role in all this? I do believe that human beings have a special role to play in God's creation—at least in our little corner of creation. Perhaps there are other beings on other planets, in other galaxies, or in other universes who are God's image bearers to their parts of creation. But having that special role doesn't have to mean that we're all God cares about in creation. In fact, in some ways creation got along better without us. Perhaps God's delight in watching stellar fusion has something to do with that— everything was going exactly the way God wanted it to go then.

This dialogue with the physical sciences was fruitful in pushing me toward a better theological understanding of time. Perhaps a better understanding of the life sciences would also lead to a better faith and to theological insights about our species. That's where we're headed next.

PART III

SPECIES

11

WHAT IS A HUMAN?

I n the 1960s, some miners in the south of Morocco uncovered a skull. It was originally thought to be Neanderthal because it is a little more elongated than our skulls are today. But over the years, several more skeletons were discovered at the same site, and the scientific consensus now is that they belong to our species, *Homo sapiens*. The fossils have been dated to about 300,000 years ago, making them the oldest known examples of our species.

The fossils are now housed in the Museum of History and

Civilizations in Rabat, Morocco. I was in Spain for a couple of weeks and discovered it would be fairly easy for me to take a day trip down to Rabat to visit the museum. I flew there but soon discovered that my total lack of Arabic and only rudimentary knowledge of French would make this trip a challenge. Through charades, loud repetition of words, and finally passing a cell phone back and forth, I thought I got my taxi driver to understand where I wanted to go. But then he dropped me off about twelve blocks from the museum. It didn't open for another forty-five minutes, according to the website, so instead of continuing the misadventure in communication, I figured I could walk.

I made it to the museum a little before it was scheduled to open at 10 a.m. and saw there was a security guard standing out front. I checked my Google translator and practiced the phrase a couple of times in French for "Does the museum open soon?" Then I walked up to him and tried saying it. I'm afraid I wasn't very convincing because he responded, "It's okay, I can speak English."

"Oh good," I said. "Is the museum opening soon?"

He replied, "No, it is closed."

"I know it's closed now," I said, wondering if his English wasn't really that good, "but it opens at 10 a.m., right? That is what I saw on the website."

"No, I'm afraid it will be closed all day today. They are doing some refurbishments." Uh-oh, if he knows the word *refurbishment*, I'm betting his English *is* pretty good. So I explained why I was in town and that I would be here for one day only. I didn't technically offer a bribe, but I asked several times if there was anything I could do, or anyone I could talk to, to get special permission to see the fossil exhibit. To his credit and integrity, he remained very firm in his refusal.

I walked away disappointed, wondering what I would do until my evening flight back to Spain, and happened upon a building whose sign I translated as "Administrative Offices for the Museums of Rabat." I thought I'd give my persuasive skills another go. After more charades and cell phone passing, the people at the front desk called a woman who came out to meet me. She too spoke English very well, so I explained the situation, telling her I was writing a book and wanted to include something about this national treasure of Morocco as a way of honoring the important scientific work here and in the country as a whole. She may have sensed I was laying it on pretty thick, and I don't know if she had any power to help me. But she got a little defensive, saying that they had posted about the closure on their Instagram account. "It would have been good," I responded with a touch of impertinence, "if it had also been posted on the actual museum website!"

"Yes," she replied, "that would have been good." In a last concession of goodwill, she took my email address and promised to contact me if anything changed. Of course I never heard from her again.

So I spent the day walking around the markets of Rabat and finally renting a table and umbrella to sit at the beach. That's not a bad way to spend a day, I suppose. But I really wanted to see those fossils.

Why? What would I learn in the museum that I couldn't by reading about the fossils? And were they even human?

Before the COVID-19 pandemic, my podcast producer, Colin, and I conducted almost all of the podcast interviews in person. The pandemic showed us that wasn't really necessary, but I have missed some of the interesting experiences that came with in-person interviews—like the one with Dr. Rick Potts.

Rick Potts is the director of the Hall of Human Origins at the Smithsonian's Museum of Natural History in Washington, DC. To record our interview, he met Colin and me at the public entrance of the museum but then took us up some back stairs. We walked through narrow hallways crowded with cabinets and drawers containing who knows what. Along the way to his office, Rick poked his head into a few rooms to greet people who were huddled over counters of important-looking artifacts.

His office was big and had stacks of papers everywhere. We sat at a table and talked about his excavations in Kenya and other places. After we were done recording, he casually handed us what I thought was just a rock that had been sitting on his desk, saying, "This is the oldest human artifact in the museum." It was an Oldowan chopper, some 2 million years old, a tool used by our ancestors to create other tools. They would strike it onto another rock and produce sharp-edged flakes that were useful for cutting.

It was an existential moment to be connected to our ancient ancestors in that way—much more meaningful than reading about it on Wikipedia (or even in this book, I'm sure!). I held in my hand a piece of their technology, which contributed significantly to the kinds of lives they led—and ultimately to the kind of people we became.

But wait a minute . . . 2 million years old and a "human" artifact? Didn't I just say the fossils in Morocco are the oldest of our species at 300,000 years?

It depends on what you mean by "human." That's a pretty important designation for my theological tradition, which asserts that we humans are the ones created in the image of God. Were those creatures 2 million years ago—not even our own species, *Homo sapiens*, but our distant ancestors—also image bearers? There seems to be some clash between the scientific terminology and our theological understanding. When did our ancestors become human? What is the difference between humans and *Homo sapiens*? What does it mean to be human?

In this part of the book, I'm addressing what I've called the "challenge of species." Briefly stated, it is very important to traditional Christian theology that human beings are created in the image of God; but evolutionary science has made it very difficult to draw clear lines between us and nonhuman creatures. Doesn't science, then, undermine the credibility of the traditional Christian story?

Part of the problem is simply linguistic. Different people use the term *human* to refer to different groups of individuals. We'll try to sort that out as we go along. The deeper problem with understanding humans in evolutionary terms, though, is that it doesn't seem like we can legitimately say that we humans are a different kind of creature. We seem only to differ by degree from other animals. How would I be able to reconcile that with my faith?

Let's start with the classification system used by scientists.

Since 1758, we contemporary humans have been classified as *Homo sapiens* in the system invented by a Swedish physician, Carl Linnaeus. In school, most of us learned the mnemonic "King Philip came over for good soup" (that's the G-rated version) to remember the hierarchy of taxa in the Linnaean system: kingdom, phylum, class, order, family, genus, species. Over time, a few other levels of classification have been added, but for our purposes here these will suffice.

Linnaeus looked primarily at body types to put individuals into these categories. He included humans, and it was something of a shock for people in his day even to appear in the same taxonomy with beasts. His response:

> *I know full well what great difference exists between man and beast when viewed from a moral point of view: man is the only creature with a rational and immortal soul. . . . If viewed, however, from the point of view of natural history and considering only the body, I can discover scarcely any mark by which man can be distinguished from the apes.*[1]

Kingdom	Animalia
Phylum	Chordata
Class	Mammalia
Order	Primates
Family	Hominidae
Genus	Homo
Species	Sapiens

Even in our own day, it continues to shock some people to hear about our relationship to other creatures. They believe we are something wholly different and want to assert, "We're not animals!"

"Well," I like to ask back, "we're not plants, right?"

In his taxonomy, Linnaeus was not explicitly making any grand metaphysical statement about what it means to be human. He was simply categorizing living things according to characteristics that are obviously shared or not shared with other creatures.

We are in the Animal kingdom because we have to eat other things (that is, we are heterotrophs). We're in the Chordates phylum because we have backbones (that is, we are vertebrates). Within this we are in the class of Mammals because we have hair, are warm-blooded, and give birth to live young. We are in the order of Primates because we have larger brains and have prioritized sight over smell. And our family is Hominidae because we have no tails or ischial callosities (calloused structures on our bums). The general characteristics used for classification have changed over time, and there are exceptions that don't quite fit the general rules—one of the most famous being the duck-billed platypus, which is a mammal that doesn't give birth to live young. But for the most part, this system of classification is still useful.

Within the family Hominidae are the other great apes, which are called "hominids." All the living hominids are separated into four genera (the plural form of *genus*) and eight species, as shown in the following diagram.

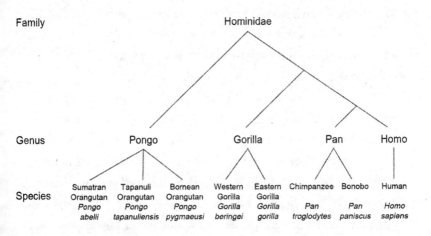

This diagram, called a "cladogram," is how scientists understand the relationship between the living hominids today. Linnaeus didn't know about gorillas or bonobos, and he had only sketchy knowledge of orangutans and chimps. Furthermore, he was working before Darwin and the theory of evolution, so he was not claiming that hominids (or the members of any family) had similar characteristics and body types because they shared relatively recent common ancestors. He just noted that their body plans are very similar.

Now we come to our genus: *Homo*. The term can be confusing, because in Greek the word *homo* means "same" (as in the words *homogeneous* and *homosexual*). But the Linnaean system uses Latin, and the Latin word *homo* means "man" in the gender-neutral sense, or what we more properly call "human being" today (like the root in our word *homicide*, which is the killing of another human being).

What does it take to be included in the genus *Homo*? That isn't entirely clear. Linnaeus didn't include criteria in his original designation, other than the cheeky phrase "Know thyself" in the place where he usually gave identifying characteristics. Perhaps it was just so obvious which individuals were to be included that it didn't need to be stated. We know ourselves, he reasoned. Our genus is the one with people in it. There wasn't a question of which species could be called human because there was obviously only one.

Then two scientific developments *did* call this into question: evolutionary science, which posits a gradual transition from one species to another without any clear line of demarcation between them, and archaeology, which discovered the fossilized remains of a bunch of other creatures that don't fit too well into the human or nonhuman categories they had back then. Today scientists recognize at least thirteen species within the genus *Homo* (ours and twelve others that have gone extinct). It hardly makes the news today when a new species is discovered and named. But the first addition to our genus was a pretty big deal. We'll meet those relatives in the next chapter.

12

OUR CLOSEST COUSINS

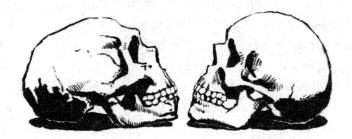

On the same trip where I didn't get to see the fossils in Morocco, I also didn't get to see a Neanderthal cave in Gibraltar. I had arranged with Dr. Clive Finlayson to get a personal tour of Gorham's Cave, where his team is excavating Neanderthal remains. But then it rained.

The cave is in the massive Rock of Gibraltar, which is mostly limestone. Limestone is soft and porous, and when weakly acidic water slowly drips through it, the rock dissolves. Over time massive caverns can form. The day before, I had hiked all the way up the rock (dodging the macaques, which are small primates that have somehow managed to become both a tourist attraction and a nuisance) and toured Saint Michael's Cave. It is large enough that they have installed an auditorium with a seating capacity of four hundred.

The next morning I went to the Gibraltar National Museum for my appointment with Finlayson. Instead of meeting him in the reception area and driving to the site as we had planned, I was ushered up to his office by an assistant. Finlayson greeted me with apologies. "The bureaucrats make these rules, and I'm afraid there is nothing to be done." Evidently, no one is allowed in the caves for forty-eight hours after it rains. They worry that the rain will suddenly dissolve something and a big rock will fall onto someone's head. Finlayson acknowledged that sort of thing happens occasionally but said, "I grew up playing in those caves even in downpours, and nothing bad ever happened." Hmm . . . I suppose if given the chance, I would have signed a waiver, donned a spelunking helmet, and risked it. But I wasn't given the chance. Instead we sat in his office and talked for an hour about his team's discoveries in the caves and what light they are shedding on Neanderthals and their way of life. It wasn't as good as being there, but it was interesting nonetheless.

It is sometimes claimed that the first known discovery of Neanderthal remains happened on that same Rock of Gibraltar in 1848.[1] Someone digging at a quarry found a skull and brought it to the attention of Lieutenant Edmund Henry Réné Flint, who presented it to the Gibraltar Scientific Society on March 3, 1848. But they didn't call it a Neanderthal, and it was assumed to have come from a human who lived there long ago—possibly before

Noah's flood, in their estimation. Nobody was suggesting it could be a different species.

But then eight years later, in August 1856, other limestone quarry miners in the Neander Valley of Germany discovered part of a skull and other large bones. These were given to a local fossil collector, Johann Carl Fuhlrott, who thought them strange and not quite human. He took the bones to Professor Hermann Schaaffhausen at the University of Bonn. These two presented their findings in June 1857 to the local Natural History Society, suggesting that the bones came from an ancient and previously unrecognized form of human.[2]

Remember, this is before Darwin published the book that so radically changed the way we think about the development of humanity. Even the realization that species could go extinct was fairly new. So it was more natural at this time to see bones like these and think they were deformed or diseased humans. A popular view among scientists was that these "Neanderthals" were simply *Homo sapiens* who had rickets, a disease that results in distortion of the bones—a claim some creationist groups continue to make today.[3]

William King holds the honor of first proposing that the Neanderthal fossil belonged to a different species. He was a geology professor in Ireland and had obtained a cast of the skull. He compared it to a vast collection of contemporary human skulls and presented his findings at the 1863 meeting of the British Association for the Advancement of Science, suggesting it be called *Homo neanderthalensis*. This remarkable claim did not generate much excitement at the meeting, though, probably because his session was held at the same time as one by Alfred Russell Wallace—the famous collaborator of Charles Darwin. But when the proceedings of the meeting came out in print, the reaction was much stronger . . . and not very supportive. Most of his peers in the field rejected his assignment of a new species name. King died in 1886, the year two nearly complete Neanderthal skeletons were discovered in Belgium. There was continuing scholarly debate into the twentieth century about how to classify these creatures, but it was increasingly accepted that they were not *Homo sapiens*.

Since the first discoveries, thousands of Neanderthal fossils have been found at hundreds of excavation sites throughout Europe, the Middle East, and as far east as Chagyrska, Russia (near the border of Mongolia). The majority of these have reliable dating from about 130,000 years ago down to their disappearance from the fossil record 20,000 to 30,000 years ago. The discoveries at Gorham's Cave in Gibraltar are among the most recent ever found, suggesting that the tip of the European continent may have been their last stand.[4]

So who were the Neanderthals, and where did they come from?

It can be frustrating to follow news about evolution in the popular press because the headlines are often misleading. When a new fossil is found, its discoverers often want to convey how important it is (which benefits their careers), and the press conveys this with hype to attract more eyeballs (which benefits their business). Too often those headlines are something like, "New Fossil Rewrites Evolutionary History" or "New Fossil Upends What Scientists Thought They Knew." As I said before, scientists argue about the details of evolutionary theory, and there are many things about our evolutionary past that we don't know. But the general picture that has emerged from the painstaking work of scientists is pretty clear and very unlikely to be completely rewritten by future discoveries. That general picture is this:

About 6 or 7 million years ago, a population of primates was separated into two subpopulations. Interbreeding between those groups stopped, and one group eventually evolved into chimpanzees and bonobos (these two separated from each other about 1 million years ago). The other group evolved into *Homo sapiens* and all the other hominins on our side of that split who have since gone extinct.

The most tantalizing quest in paleoanthropology is to reconstruct the evolutionary history of the hominins. I said in chapter 2 that the Hall of Human Origins at the Smithsonian says there are fossil remains from more than five thousand individual hominins. Where do they fit on our ancestral family tree?

Currently six different genera are recognized since the split with the chimpanzee lineage. In the past, scientists tended to see each new discovery as part of one evolutionary lineage, as though every ancient fossil found were a direct ancestor of ours. Now scientists are pretty confident that our family tree is bushier, with lots of offshoots that didn't lead to us.

Think of it like this: My ancestors immigrated to the United States primarily from Germany and Switzerland. But if you go to a centuries-old cemetery in Darmstadt, Germany, the birth city of my eighth great-grandfather, Hans Christoph Stumpf, it would be pretty remarkable if you picked out one tombstone at random and hit upon a direct ancestor of mine. Many of the "fossils" in that cemetery would be fairly close relatives of mine (much closer than those found at a cemetery in Japan), but they represent different lineages that did not lead to me.

So too with any hominin fossil we find. Of course there is an unbroken ancestral line from 6 million years ago to us today. But scientists now believe there were many other branches of the ancestral line that eventually went extinct. Any fossil we find, then, has a much greater chance of being on a dead-end branch rather than being one of our direct ancestors. Still, any hominin fossil will be more closely related to us than specimens from other family trees, and they tell us something about our development.

Fossils assigned to the genus *Ardipithecus* are dated from 4.5 to 5.5 million years ago. They show the first signs of walking upright on two legs but still have an opposable big toe for grasping tree branches.

Around 4 million years ago, the grasping big toe was lost, and a more modern foot developed. That suggests they spent little time in the trees and primarily walked upright on the ground. Several species in the genus *Australopithecus* lived simultaneously for a couple of million years, including the most famous fossil find, Lucy, an *Australopithecus afarensis*, dated to 3.2 million years ago. Scientists' best guess now is that one of the species of *Australopithecus* eventually evolved into our genus, *Homo*.

More than 2 million years ago, *Homo erectus* evolved in Africa and then became the first hominin to leave the continent. Fossils of this species have been found from Central Europe to China. They had bodily proportions similar to ours, with longer legs and shorter arms. That probably meant the end of any significant time in the trees. *Homo erectus* was highly adaptable and survived for almost 2 million years, finally passing out of the fossil record about 110,000 years ago—which means they overlapped with our species.

Homo heidelbergensis may have evolved from a *Homo erectus* population that remained in Africa. It seems that about 500,000 years ago, some of them went up from Africa into the Middle East and Europe, and this is probably where the Neanderthals came from. *Homo neanderthalensis* are recognizable in the fossil record there starting 200,000 to 300,000 years ago and were spread across Europe and parts of Asia by 130,000 years ago.

Back in Africa, some of the *Homo heidelbergensis* population evolved into *Homo sapiens* over that same time period. Some of them began to leave Africa in sizable numbers around 50,000 years ago. There are very few Neanderthal sites less than 40,000 years old. It surely isn't coincidental that they disappeared as we arrived.

What must that encounter have been like? Did *Homo sapiens* see something of themselves in the Neanderthals they met?

They looked a little different. Neanderthal skulls have a strong ridge above the eyes, they are longer and lower than ours, and they don't have much of a chin. These might not sound like huge differences, and they might be thought attributable to normal variation. That's what Thomas Huxley claimed in 1863, just a few years after Darwin's book was published. Huxley was one of Darwin's chief defenders (often called "Darwin's bulldog"), and so he might have been expected to see these skulls as a missing link of sorts between humans and beasts. But the fact that the brain volume of the Neanderthal skull was about the same as ours led him to conclude it was just an extreme variation of a *Homo sapiens* skull.[5]

Truth be told, according to the data we have now, the average volume of Neanderthal brains (1,410 cubic centimeters) was slightly larger than ours today (1,350 cubic centimeters).[6] But when it comes to brains, size isn't everything (sperm whale brains are about 8,000 cubic centimeters!). More important is which regions of the brain are included.

Part of that is shape, and the difference in overall skull structure between *Homo neanderthalensis* and *Homo sapiens* is much greater than the difference between species as clearly distinct as lions and tigers.[7] Neanderthal skulls are more like footballs, while ours are more like basketballs. That gives us a lot more space up front to form a bigger frontal lobe and neocortex, which are where more advanced thinking takes place.

Even at the molecular level, it looks like Neanderthals had far fewer of the genes that have been recognized to contribute to self-awareness.[8] Just before I went to see Dr. Finlayson, a new study came out claiming to show that a very slight difference in one of our genes makes neurons grow faster in the frontal lobe of our brains than they did in Neanderthals. I asked Finlayson if he had seen this. "I know of it," he said, "but haven't gotten around to reading it. Sounds to me like another wild goose chase trying to find the missing link."

Finlayson is one of the leaders of the school of thought that says Neanderthals were just as advanced as our *Homo sapiens* ancestors. In the early days of Neanderthal fossils, it was assumed that they were brutish cave dwellers. The pendulum has swung the other way among many researchers now. There is clear evidence that Neanderthals controlled fire, adorned themselves with shells and feathers, and buried their dead. Of course they didn't speak our languages or have the technology and culture we do today. But neither did the *Homo sapiens* living 50,000 years ago.

When Finlayson was laying out the case to me that Neanderthals could do anything that *Homo sapiens* did back then, he stopped and said, "I won't mention names, but there is a respected American paleoanthropologist who is an advocate for modern human superiority. At least once he said to me, 'Clive, you're the defense lawyer for Neanderthals.'"

"I bet I can guess who that is," I ventured. I had read enough on the topic to have a pretty good idea. "Ian Tattersall?" I asked.

"Yes, of course it's him, but I take that label with pride! Don't worry. We get on very well. We're good friends."

Tattersall is a curator emeritus at the American Museum of Natural History in New York. I had dinner with him once at a workshop BioLogos hosted, and I found him to be very engaging. He's written lots of books about the evolution of our species, and he says the big question about Neanderthals is whether they had the kind of symbolic language that we do. He doesn't think the evidence shows that they did. In spite of that, he admits they were very smart and had "the most sophisticated expression on record of the ancestral, intuitive style of hominid cognition." But our symbolic reasoning is qualitatively different from the way other animals process information. Our words are not simply cues that elicit responses; instead, they are abstract symbols that mean something within a community. That gives our communication and reasoning tremendous advantages. Our symbolic language is adaptable to new situations that have not been encountered before; it is recursive, or able to form much more complex ideas that are nested within each other, like "I believe that you believe that I believe something you don't believe"; and it allows mental time travel to talk about the past and future. If Tattersall is right, these abilities developed after our separation from the Neanderthal line.[9] He states, "It seems most reasonable to conclude that, while a bright Neanderthal occasionally did something intriguing that anticipated what a fully symbolic human might do, such expressions were not a routine part of the Neanderthal behavioral repertoire."[10]

So Neanderthals were pretty close to us, but probably not quite the same as us. But were they human?

It is easy to say whether a particular individual is human or not when the next closest thing is a chimpanzee. Chimps may be clever, but none of them are writing books, hosting podcasts, or sending astronauts into outer space. However, when you add in Neanderthals, Heidelbergensis, Erectus, Australopithecus, Ardipithecus, and the rest of the hominins, the transition between chimps and us looks a lot smoother.

A smooth transition from beasts to humans is difficult to understand in theological terms. But a lot of time helps. In the next chapter, we'll see how much time we're talking about.

OUR ANCESTRY IN
BASEBALL CARDS

As a kid, I collected baseball cards. Not the kind you see today that sell for thousands of dollars on eBay and are sealed behind glass. I rode my bike to the drugstore and paid a quarter for a pack of twelve Topps cards and a stick of gum. I'd open them on the spot, find one from a team I disliked, and put it in the spokes of my bicycle wheel for the ride home so I sounded like a motorcycle (or so my friends and I thought).

I organized my cards in a big shoebox, sometimes by teams, sometimes by player position. I still have that box of cards, and I've always been amazed by how many can fit inside. So when I read in a Richard Dawkins book about stacking up postcards to represent the generations on our family tree,[1] I naturally applied the illustration to my baseball cards.

Imagine that instead of baseball players, the cards each have a picture of a generation of your ancestors. In my stack of cards, for example, I'm on the top card, then my parents are on the next one down, my grandparents are on the one below that, and so on. According to a family friend who dabbles in genealogy, the thirty-fifth card in my stack would have on it Olaf II, the patron saint and king of Norway back around the year 1000. (Don't be too impressed with my royal lineage; almost all of us are descended from prominent people that long ago who did pretty well at passing on their genes!) In my case, that works out to an average of about thirty years per generation, but of course this will vary somewhat for others. Scientists commonly use twenty-five years as the average length of a generation for humans. These days women tend to give birth later in life; they used to do so at a younger age. And men have typically been older than women when their children are born. But over long stretches of time, these numbers average out to around twenty-five years per generation.

I found my big box of baseball cards down in the basement storage room and took out a stack of the crispest ones (not the ones that had been in the bicycle spokes) to measure how thick they are. I stacked them up an inch high and counted fifty cards. Using the average of twenty-five years per generation, a one-inch stack almost takes us back to the year 770, when Charlemagne would appear for anyone with European ancestry. But remember, the number of ancestors you have doubles with each generation. By the time you go back fifty cards to your forty-seventh great-grandparents, Charlemagne is only one of more than 500 trillion of them! But that's more people than were alive then (or who have lived in all of human history). How does that work?

When someone marries a relative, that reduces the number of ancestors their offspring have. For example, Charles Darwin married his first cousin, Emma Wedgwood. Charles's mother (Susannah Wedgwood Darwin) and Emma's father (Josiah Wedgwood II) were brother and sister with the same parents. So their children only had six great-grandparents instead of eight. By the time you go eight or ten generations back, most of your ancestors are distant cousins of each other, drastically reducing the number of unique ancestors you have.

Let's disregard the total number of people pictured on any card and just think about how long ago various individuals lived.

Say your stack of baseball cards is now about 3.25 inches tall. You're still on the top, your parents next, and so on. The stack of cards isn't that tall—only about half the height of most smartphones— but the bottom card of that stack might very well have the biblical Abraham on it. He probably lived around 2000 BCE, or 161 generations ago.

In the previous chapter, I mentioned that some *Homo sapiens* started moving up into the Middle East and Europe about 50,000 years ago. They encountered Neanderthals whose ancestors had been living there in reproductive isolation from *Homo sapiens* for several hundred thousand years. That much time gave rise to enough differences that most scientists today consider them a different species. But that's just the blink of an eye in evolutionary time, and they were still compatible enough with *Homo sapiens* that they could interbreed. All of us today who have ancestry from Europe or Asia are 2 to 4 percent Neanderthal, according to our DNA. This means that around two thousand generations ago, one of your 1,997th great-grandparents was a Neanderthal. In our baseball card thought experiment, a Neanderthal appears at the bottom of a forty-inch stack.

What about the beginning of our species, *Homo sapiens*? I said before that the fossils in Morocco are thought to be about 300,000 years old and very near the beginning of our species. That equates to about 12,000 generations, or a stack of baseball cards twenty feet high.

Let's go back even further to the beginning of our genus, *Homo*. That's closer to 3 million years ago—ten times further back than our species' origin. So there are 120,000 generations between us and them, and a stack of baseball cards two hundred feet tall!

It's astounding to think about all the people who have lived since then. Again, don't think that you have trillions and trillions of 199,997th great-grandparents during that first generation of *Homo sapiens*. (Nor should you think there actually was a first generation of *Homo sapiens* in the sense that their parents were a different species. That's not how evolution works.) Estimates from our genetic diversity today can be made about the total number of *Homo sapiens* that have ever lived. The best, most rigorous analysis I could find of this was from the Population Reference Bureau in Washington, DC. They give a total number of 117 billion.[2] But they were only taking *Homo sapiens* back to 190,000 BCE. I think we can safely add a few more billion going back another 110,000 years to 300,000 before present, the date now accepted for the emergence of our species. If so, the total number of *Homo sapiens* who have ever lived is about 120 billion.

Were they all humans? The relationship between human and *Homo sapiens* is a tricky one. Lots of people in popular discourse use the terms *human* and *Homo sapiens* synonymously: everyone who is a human is also *Homo sapiens*, and vice versa. But it is more common among scientists to treat *human* as naming a larger category that includes all the other species within the genus *Homo*: *neanderthalensis*, *heidelbergensis*, *sapiens*, and all the rest. Remember that *Homo* literally means "human" in Latin. According to that usage, we could say all *Homo sapiens* are human, but not all humans are *Homo sapiens*. (If that's confusing, think of it as the same as saying all fish are animals, but not all animals are fish.)

Some people wonder if the nesting relationship between the terms should work the other way, and *human* should be more restrictive than *Homo sapiens*. They suggest that all humans are *Homo sapiens*, but not all *Homo sapiens* are human. One way of doing this is scientifically, by acknowledging subspecies. That classification is sometimes called *Homo sapiens sapiens* and refers just to those who are both anatomically human and are also "behaviorally human." The way our anatomy looks today was in place by 300,000 years ago, with only very minor differences. The *Homo sapiens* then were anatomically human. But some anthropologists think there was a massive shift in behavior about 50,000 to 60,000 years ago. That is when we get the first definitive evidence of things like art, rituals, and other symbolic behaviors. (Some anthropologists think those behaviors began much earlier but were not preserved in the fossil record.)

People who prefer this option might say that it is only proper to call individuals "human" once they began acting the way we do. That means there could have been lots of *Homo sapiens* (all those before about 60,000 years ago) who were not human.

There could also be a theological justification for restricting *human* to a subset of *Homo sapiens*. You might say that humans are the ones God entered into a different relationship with, designating them as image bearers to the rest of creation. That could have happened at some point during the evolution of *Homo sapiens*, perhaps after they became behaviorally modern (or perhaps just before, and that was the cause of our becoming behaviorally modern). Theologians don't all agree on this, though. In fact some prefer the option of letting *human*—and therefore *image bearer*—be the larger, more inclusive category.[3]

The resolution of how we use *human* is not going to come from a dictionary definition or from more careful scientific examination of fossils. Language usage is fundamentally a social construction, and the rules of language are determined by how people actually use the words over time. We find ourselves at a point in time where not everyone is using *human* to refer to the same individuals. We shouldn't let that ambiguity conceal a deeper truth from my illustration about the stack of baseball cards. Let's go back to it.

What if we could trace our family tree all the way back to the last common ancestors we had with chimpanzees and bonobos? The estimates here start to get a bit fuzzier. In years, the split would have occurred about 6 or 7 million years ago. The average length of a generation over this time was probably shorter than twenty-five years. Female chimps today can start having babies at about fourteen years old, and males reach sexual maturity at about sixteen years old. If we manipulate the assumptions a bit (while staying well within the fuzzy values), we get some interesting numbers. Say the most recent common ancestor of humans and chimpanzees lived 6.5 million years ago and that the average generation over that time was 19.5 years. That gives us 333,333 generations—one-third of a million. And a stack of that many baseball cards is 555 feet tall, which is the height of the Washington Monument!

I really like the image of a stack of baseball cards that high. First, it conveys clearly our separation from our common ancestors with any other species living today. It's a really, really long time to go back through all those generations. Using the numbers above, chimpanzees today are our 333,334th cousins (about 20,000 times removed because of their shorter generation time). These are not close enough relatives that you need to send them Christmas cards!

Second, the stack of baseball cards shows us how gradual the changes are. Looking from one card to the next in that stack, you wouldn't notice any substantial differences between successive generations. They might have different colored hair or other features different from their parents or kids—just like we do today—but you wouldn't ever say that there was a species change between one card and the next.

The cutoff for when a species changes over from one to the next during evolutionary history can't be made at the level of individuals. We know that 500,000 years ago there weren't any *Homo sapiens*. But we know there were creatures living then who had babies, who had babies, who had babies, and so on, and that there are members of that ancestral line 100,000 years ago who are rightly called *Homo sapiens*. There is no precise point at which one species turns into another.

If you prefer logic, we could say that the relation "is the same species as" is not transitive. A transitive relation is the following: if x has the relation to y, and y has the relation to z, then x will have the relation to z. For example, the relation "is taller than" is transitive: if Shaquille O'Neal is taller than Taylor Swift, and Taylor Swift is taller than Tom Cruise; then we know Shaq is taller than Tom Cruise without having to put them back-to-back and confirm. Most of the relations we use to compare things are transitive. But not so with "is the same species as": just because x is the same species as y, and y is the same species as z, that doesn't mean that x will be the same species as z (at least not if there are a bunch of generations between x and z).

It seems, then, that all we will ever get from evolution is small changes by degree. How are we supposed to reconcile that with the theological commitment that we humans are a different *kind* of thing?

DEGREES AND KINDS
IN THE CAVES

When we see the continuity of *Homo sapiens* with other species, it's natural to wonder whether there is anything special
about us. The changes we could observe from one generation to the
next in the stack of baseball cards are almost imperceptible. Does
that mean we differ from Neanderthals or chimpanzees only by degrees? Or can we legitimately say we humans differ in kind from
everything else?

Darwin's most famous book, *On the Origin of Species*, did not explicitly address the ancestry of humans. He worried how people would react to being told that they were related to beasts. But the implication was obviously there, and twelve years after its publication he laid it all out in another book, *The Descent of Man*. He didn't know all the specifics of our family tree, including which other creatures are most closely related to us or when ancient populations split into different species. Scientists today have vastly more evidence available to them and are still debating these kinds of questions. But Darwin got the general idea right: the species we see today—including *Homo sapiens*—are descended from common ancestors in the past. That must mean, he reasoned, that all the differences we see between us and the beasts are simply differences of degree. Here is a characteristic passage that summarizes his view:

> *There can be no doubt that the difference between the mind of the lowest man and that of the highest animal is immense. . . . Nevertheless, the difference in mind between man and the higher animals, great as it is, certainly is one of degree and not of kind. We have seen that the senses and intuitions, the various emotions and faculties, such as love, memory, attention, curiosity, imitation, reason, &c., of which man boasts, may be found in an incipient, or even sometimes in a well-developed condition, in the lower animals.*[1]

According to Darwin, the impressive capabilities of humans are nothing more than small changes in degree over long stretches of time. That means we're not a different kind of thing. But not everyone agrees.

Back in chapter 10, I quoted a passage from G. K. Chesterton about God's childlike exuberance in watching the sun rise every day. In another of his books, *The Everlasting Man*, Chesterton reflects on the conception his culture had of "cavemen" as unthinking brutes. But when he was writing in 1925, the only hard evidence of what life was actually like inside the caves tens of thousands of years ago was some recently discovered cave paintings. These, Chesterton claims, were not feeble attempts at art by unthinking brutes. Rather, he argues, we cannot but recognize them as the work of people like us, and the capacity to create art puts us in a different class than all other animals:

> *It must seem at least odd that [one] could not find any trace of the beginning of any arts among any animals. That is the simplest lesson to learn in the cavern of the coloured pictures; only it is too simple to be learnt. It is the simple truth that man does differ from the brutes in kind and not in degree; and the proof of it is here; that it sounds like a truism to say that the most primitive man drew a picture of a monkey and that it sounds like a joke to say that the most intelligent monkey drew a picture of a man. Something of division and disproportion has appeared; and it is unique.[2]*

Who is right? Darwin or Chesterton? Do we only differ in degree from other animals? Or are we a different kind of creature altogether? And when we apply it back to the theological question of being image bearers of God, does it matter?

To help me answer these questions, I wanted to see some of this cave art myself. I went to Bordeaux, France, and rented a car. From there I drove a couple of hours east along a major road and then down into the remote countryside to a picturesque inn. The next morning I drove further out into the countryside, and my GPS led me astray. (This was the same trip as the swing and a miss at seeing cool things in Morocco and Gibraltar. Was everything colluding to keep me from achieving my goal?!) Eventually I made it to Rouffignac Cave. Its existence has been known for centuries, and it was used during World War II as a hideout for the French Resistance. But it was only in 1956 that the cave art deep inside was rediscovered and made known to the public. Some of the passageways were dug out, and an electric tram was put in for visitors.

I bought my ticket and even coughed up an extra five euros for an old iPod Touch the clerk pointed at, saying, "English." I had to wait with other visitors in the outermost chamber of the cave while a school group took the tour first. During this time I discovered that the English on my iPod was simply someone reading the English captions of pictures that were already on display there. I wasn't very impressed.

Finally about twenty of us were let through a door leading deeper into the cave, where we boarded the tram. It felt like a ride at Disney World (it's a small world after all!), and my expectations were not very high. I could only catch about one out of ten words from the French guide as we descended more than a kilometer farther into the cave. He pointed out some scratches on the wall, which I had learned from the pictures in the waiting room were probably from ancient bears that once hibernated in the cave. I guess that's kind of cool, and the other visitors were having a good time chatting in other languages and laughing as they pantomimed sleepy bears scratching the walls. So far this excursion wasn't really helping me in my quest to understand what our ancient ancestors were like, though it did provide some insight into my fellow travelers!

After fifteen or twenty minutes, we reached a larger chamber and the tram stopped. The guide said something and everyone else got out, so I followed along. Then some lights came on, and all the chatting stopped. A collective gasp of amazement issued from our group as if choreographed. It was understandable in any language. We stood in the midst of a herd of mammoths that seemed to leap from the cave walls.

There are at least 255 figures drawn in the cave, most of them mammoths, which hardly appear in other ancient cave art. These are not random scratches on the wall, but gracefully drawn figures incorporating the contours of the cave wall. They were planned out with intention and executed with skill. The difference between them and what the hibernating bears left behind is shockingly obvious.

The dating is tricky, but the best estimates are that the mammoths of Rouffignac were drawn around 13,000 years ago. It was really remarkable to stand there in the space where our ancestors stood so long ago. They traversed the kilometer down into the cave without an electric tram, through narrow passages lit only by a torch that burned reindeer fat. That couldn't have been easy and made the difficulties I went through to get there seem trivial. Why did they do this? Why had I done this? To create art. To see art. There was no survival advantage conferred by clambering deep into a cave to draw. In fact, it was probably pretty risky. But wow, did it leave an impression. They were like us.

Hoping that my fortunes were changing and that the trip was worth all the hassle, I decided to drive another thirty minutes to the much more famous cave painting site of Lascaux. There the original is closed to the public, but a series of replicas has been built. Lascaux IV is an almost perfect reproduction of the original cave. I took a tour (in English!) that guided us through the cave and its amazing artistic portrayals of animals. From the monochromatic line drawings of Rouffignac, Lascaux is like stepping into technicolor. And even though this was a replica, it wasn't hard at all to believe I had stepped into a prehistoric world and seen the humanity of the artists.

Seeing these sites, I couldn't help but agree with Chesterton: "Something of division and disproportion has appeared; and it is unique." These people were us. It's a small world after all.

But Darwin wasn't entirely wrong about the nature of humans with respect to other animals. He was surely right about the fact that we share common ancestors with the other life-forms on Earth. Chesterton almost gives you the feeling that he thinks something magical happened to give those ancient humans the abilities they had. Darwin's point—and what his successors have plausibly shown—is that those abilities really can be achieved by a series of adaptations.

The tricky and often misunderstood part of the claim that we differ in kind from other species is that the capacities that so clearly set us apart (art, language, morality, culture, and so on) didn't spring from nowhere. They are dependent on other components of behavior and on our brain structures, and those things *do* have evolutionary stories and *did* develop by degrees. So we can find hints or precursors of our distinctive abilities in other species, and the answer to the question of whether we differ in kind or only by degrees seems to be . . . yes.

I was going to need a way to better unpack and understand this "both/and" answer. I found it among the trees.

FINDING DEEPER FAITH
AMONG THE REDWOODS

My wife and I were recently in San Francisco, and I wanted to see some redwoods. My last name has always given me an affinity for trees (and something of an inferiority complex!), so we made a reservation to visit Muir Woods National Monument, which is just a little north of the Golden Gate Bridge. It is one of the few places where old-growth coastal redwoods are still standing and relatively accessible.

After an atmospheric river washed out our plans on the first try (and fearing a return of my bad luck), it looked like the next day would work. We drove across the bridge, and the lingering rain and fog gave the whole experience a fairy-tale quality. Descending into the valley where the park is located, we seemed to be entering Rivendell or Lothlórien—the Elven forests in Tolkien's *Lord of the Rings*.

We walked the path that snakes along the creek where the redwoods grow, marveling at the incredible height and girth of the trees. One section of the park, where some of the largest and oldest trees have stood for more than a thousand years, is even called Cathedral Grove. A kind of reverence naturally settled in, and I imagined God almost bragging a little bit, whispering to us, "See these things I made? Pretty impressive, huh!" It was almost like they were a different kind of thing than the maple trees from my hometown.

For the podcast, Colin and I made a series of episodes on what it means to be human. We interviewed a number of experts from different disciplines and attempted to incorporate their insights into an extended audio essay on the topic. One of those experts was Jeff Schloss, a biologist at Westmont College.[1] He gave an illustration that lets us see the "both/and" answer to the question of whether we differ in kind or degree from other animals. He asked us to imagine that all the plants in the world are only the size of rosebushes, and then you come across a giant redwood tree. Wouldn't most people think that the redwoods are a completely different kind of thing?

Schloss went on to note that some of his evolutionary biologist colleagues have called humans "spectacular outliers" in this respect. To call us an "outlier" means that we're somehow measuring similar qualities—outliers are on the same graph as the rest of the data. So the difference *can* be characterized as one of degree. But when that degree of difference puts you so far off the chart from everything else that has a little bit of that quality, it may merit being called a difference in kind. If walking through redwoods that can reach three hundred feet high made me feel that difference with respect to other trees, how much more would I feel it if the only other plants were three-foot-high rosebushes?

On the surface, the discussion about other species and what it means to be human is a clash between the scientific terminology of evolution and the language I inherited from my faith tradition. On the one hand, the Bible only says that humans were created in the image of God, and therefore it has become part of the theological tradition that we have been set apart from the rest of creation; we're a different kind of thing. On the other hand, evolution makes it pretty clear that we are continuous with the rest of creation—only different in degree; it's hard to determine conclusively which individuals over the last several million years should count as humans. To form a coherent view of the world that takes into account both of these perspectives, something has to give. The theological and scientific claims can't each be seen as absolute. They must be seen as dialogue partners who come to the topic from different perspectives, trying to sort out the truth behind their limited experience. A couple of insights from my journey of investigating what a species is help overcome this challenge and bring these perspectives together.

Like most of these debates, a resolution depends significantly on the definitions of the words. What does it mean to be a different *kind* of thing as opposed to merely differing *by degree*? The difficulty is that for any individuals, we can find some ways they are the same, and some ways they are different. The early Christian theologians had the same difficulty in talking about the nature of Christ. Was he human or divine? They wanted to claim that he was both in some sense, though some leaned more heavily in one direction or the other—and were usually accused of heresy by the other side! The different camps were not all-or-nothing positions, and instead might be placed along a continuum (not unlike the way evolution works). Perhaps it is helpful, then, to talk about a spectrum of positions.

At one end of the spectrum is the position that humans are just another animal, vertebrate, mammal, primate, and hominid—they differ only by degrees from the others in those categories. At the other end of the spectrum is the position that humans are completely different and set apart from all other life—a difference in kind. There are lots of intermediate positions between these two extremes, as people try to capture the obvious truth that there is at least some difference and some continuity between us and other animals. I'll give labels to several of these positions, knowing full well that the terminology is not completely standardized.

One cluster of positions closer to the first end of the spectrum might only want to refer to human distinction. Yes, humans or *Homo sapiens* are different from other species. But every species is different from other species—that's what it means to be a species. People who are drawn to this categorization are content to detail what it is that makes us distinct from everything else. They could do that same sort of analysis with any other species because, again, that's what it means to be a species—we have a set of characteristics that is distinct from that of other creatures. The fact that we are unique in some identifiable ways doesn't set us apart because everything else is unique too.

At the other end of the spectrum are people who prefer to talk about human exceptionalism. This has the connotation that we don't just have different characteristics and abilities the way everything else does, but that these differences are better and more valuable to God. Human exceptionalism is a very anthropocentric view of the created order (meaning we regard humankind as the most important element of existence). It seems to suggest (and some of its proponents claim outright) that everything else in creation exists for us.

Both of these extremes seem to be missing something important—the equivalent of the Docetists who emphasized Christ's deity to the exclusion of his humanity, and the Arians who emphasized Christ's humanity to the exclusion of his deity. The resolution of these two tendencies seems to some people to have been accomplished simply through assertion: Jesus is fully human and fully divine. But pretty sophisticated thinking in the Christian tradition and careful attention to language made the assertion credible. I wonder if there is a way to do something similar with the two tendencies about our species' nature.

In the previous chapter, I explained how I became convinced we're not simply different in degree from other animals. But neither am I comfortable with the extreme forms of human exceptionalism. One of the lessons of part I was that it's not all about us. If God can spend billions of years just creating heavy elements through stellar fusion, then I don't see any way around admitting that God must enjoy these other facets of creation and value other creatures for more than their instrumental value to humans. But that doesn't mean we're no different from everything else.

A few years ago, I compiled some quotations from contemporary researchers in diverse fields who are engaging with the question of what it means to be human.[2] None of them have religious motivations for their conclusions (so far as I know) but rather are presenting the results of empirical research. Here are a few examples:

> Human societies represent a huge anomaly in the animal world.
>
> —*Ernst Fehr and Urs Fischbacher, economists*[3]

> A hundred years of intensive research has established beyond reasonable doubt what most human beings have intuited all along; the gap is real. In a number of key dimensions, particularly the social realm, human cognition vastly outstrips that of even the cleverest nonhuman primates.
>
> —*Kevin N. Laland, evolutionary biologist*[4]

> Humans woke up from being organisms to being something quite different: embodied subjects, self-aware and other-aware in a manner and to a degree not approached by other animals.
>
> —*Raymond Tallis, neuroscientist and philosopher*[5]

> That difference is indeed sufficiently striking to make human life radically different and to furnish us with such unique dignity as we actually have.
>
> —*Mary Midgley, philosopher*[6]

These quotes all seem to agree with G. K. Chesterton's reflection on the ancient people who painted in caves: "Something of division and disproportion has appeared; and it is unique."

So how do we articulate a position in the middle of the spectrum where we are more than distinct but maybe not exceptional in the sense that it's all about us? Let's first call that position *human uniqueness*. But perhaps we can't stop there, as that sounds like a synonym of *distinct*. So some people have taken to saying that humans are "*uniquely* unique."[7]

Yes, humans are different, but so is everything else. Eagles have better sight, gorillas are stronger, redwoods are taller, etc. But the claim here is that humans are unique in a different kind of way. We're not simply noting one set of abilities that we do better than anything else. It's like we're taking the analysis to a different level: not just looking at specific differences between species but looking at how much those differences differ from each other. And on that measure, humans are radically different.

I don't go all the way to exceptionalism, though, because there continues to be a scientific case that even our radically different kind of existence has evolved through small steps. That helps me understand how we might be called to be something different within the created order, even though we share common ancestry with everything else.

There is another insight that helps bring the scientific and theological perspectives on humanity together: God didn't create things originally the way they were ultimately intended to be. That may not sound too profound, but I think it has not been widely recognized and appreciated. The typical characterization of creation history is that everything was perfect before Adam and Eve sinned, as though its initial state was the way God always wanted it to be. But we can see from the Bible itself that this is not a very faithful reading of scripture. (We'll explore this further in chapter 22.)

Starting from the first chapter of Genesis, right after humans were created, God says to them, "Be fruitful and multiply and fill the earth and subdue it" (Gen. 1:28). It seems that God wanted the Earth filled and subdued, but that was not the way it was created. Why not?

We might also look to salvation history in the Christian tradition. It seems God initiated a *process* through which all people would be reconciled to God: first by calling Abraham, through whom all nations would be blessed (Gen. 22:18), and then by becoming incarnate in Jesus Christ. It doesn't seem right to say that Jesus was God's "plan B." Rather, God planned from eternity past to become human. That didn't happen right away, but only in the fullness of time. Why?

Or finally and ultimately, if it is God's plan to create an eternal heavenly kingdom populated with people who worship God, why not just start with that in the first place? The answers to all these questions can only be speculative, but they seem to me to point in the direction of *valuing the process* and wanting to give space to other agents to express their own freedom and creativity. God is the Creator, not the dictator.

With this understanding, we can make sense of the otherwise seemingly competing claims that God intended to create human beings, but *Homo sapiens* were late in emerging in the history of the universe. God didn't initially create the universe the way it was ultimately intended to be. Furthermore, this kind of thinking helps resolve a problem that I had in understanding God's continued activity. If everything had been initially created in its final perfect state, then God wouldn't have much to do but sit back and watch. That's the God of deism rather than the God of Christianity, who is actively involved in how things develop over time.

Finally, though here I can only speculate, it seems more than a tad human-centric to think the present time is the culmination of the history of the universe. When we acknowledge that God didn't create things initially the way they are ultimately intended to be, we can't just assume that the way they are now is the way they are ultimately intended to be—that we are the last stop on the journey of creation. God seems to be up to something more than we have seen so far. In the conclusion to the book, I'll speculate a bit on where things might be headed.

So far I've come to understand how the Bible could be written by ancient people who had very different scientific knowledge than we do and yet could be used by God to convey important spiritual truths that are relevant to us all. Then I grappled with the incredibly long stretch of time our universe has existed and the relatively miniscule amount of time we humans have been on stage for that entire drama. That forced me to rethink some of God's priorities, but ultimately let me see God's creativity and delight in creation in a deeper way. And in this part, I found that science's perspective that we are continuous with other species, and theology's perspective that we are something different, are both valid and ought to be held in tension.

We have two more challenges to face . . . and they keep getting harder. The next one is the challenge of the soul. We'll confront in greater detail—both scientifically and theologically—the difference between us and other creatures, and we'll wonder how that sort of difference could evolve.

PART IV

———

SOUL

WHAT HAPPENED TO THE SOUL?

One of my favorite movies is *O Brother, Where Art Thou?* It's a retelling of Homer's *Odyssey* set in the Deep South in the middle of the Great Depression. In one scene, the main characters Everett, Pete, and Delmar are driving through the countryside when they stop to pick up a hitchhiker named Tommy who is carrying a guitar case. When they ask what he's doing out in the middle of nowhere, he says that he had to be at the crossroads last night at midnight to sell his soul to the devil so he could learn to play his guitar real good. Delmar laments, "Oh, son. For that you traded your everlasting soul?" Tommy replies, "Well, I wasn't using it."

Besides being pretty funny, Tommy's response sums up a common attitude toward souls: we might still have them, but we don't think they do anything. In that vein, I have an album by a local musician called *Vestigial Soul*. Have our souls become like an appendix or wisdom teeth—relics of a bygone era that have lost their purpose for today?

In this part of the book I'm confronting what I've called the challenge of the soul. Part of the problem here is terminology. For lots of people like Delmar, the word *soul* has a distinctively religious connotation about the afterlife, as though that is its only function. I'll speculate a bit about an afterlife in the conclusion to the book, but here I'm more interested in the soul as "the real me" or "the self" and how it affects our experience in the here and now.

We are not mere objects that are acted on by outside forces. We are also agents or subjects who are capable of acting because of reasons. We have a perspective on the world, we experience things through a distinct center of consciousness, we can think and reflect on our experience, and we can choose how to respond. The word *mind* is often used to refer to some aspects of this, but *soul* is a more encompassing term for the transcendent aspect of our existence that makes us more than machines.

So the challenge of the soul I'm addressing is this: how could the soul evolve?

For starters, think back to the example from chapter 8 about the boiling teakettle. We saw there are two very different ways of explaining the event—one that appeals to the scientific details of causes, and the other that gives a personal story about reasons.

Instead of a boiling teakettle, what if we ourselves are the things we are trying to explain? Are there still two different, legitimate ways of describing us: a scientific one and a personal one? I think there are, depending on the context and circumstances.

Our bodies, including our brains, are made up of material particles—the same kinds of particles that make up moon rocks, sycamore trees, and polar bears. If we could zoom in close enough to any object we encounter, we'd see atoms of carbon, hydrogen, oxygen, and other naturally occurring elements. For the objects that are our bodies, modern science and medicine have gotten very good at discovering and treating problems because they increasingly understand how our bodies work as part of the natural world. It is perfectly legitimate to use the language of science and causes to explain the operations of these objects—our bodies.

But does the language of science tell the whole story about human beings and what we are? Can we say more about us than what our physical components are doing? We have beliefs that are true and false. We have free will and are morally responsible. We recognize beauty and produce art. These are not the kinds of qualities that can be ascribed to atoms. The true, the good, and the beautiful transcend the physical realm somehow, and we are part of that transcendent world. So we are not just objects in the natural world, but we are also subjects—agents capable of acting for reasons. The things we do are not merely the necessary consequences of prior causes.

We have to use both ways of describing humans to do justice to the range of our experiences. But it is not as if these are separate from each other. I can decide to do something with my body, and it carries out my instructions (though admittedly less so the older I get!). And the direction of influence goes the other way too: I might have a headache or not get enough sleep, and that directly influences my conscious experience or how I treat other people. There are even examples of brain tumors that seemed to cause people to commit morally reprehensible acts.[1] The two aspects of ourselves are clearly connected, but they also seem to be two quite different perspectives on a person.

We are objects, and we are subjects. We are body and soul. Acknowledging this is the first stage in addressing the challenge of the soul. The next step is to consider some possible explanations for why we seem to have these two perspectives.

One way of explaining why we are both an object and a subject is to say that we are actually made up of two different kinds of substance—one material substance (the body) and one immaterial substance (the soul). The usual poster child for this view is René Descartes. He was a seventeenth-century philosopher who coined what has become the most famous (and most parodied) philosophical quip, "I think, therefore I am." With this short argument, he claimed that even if he was mistaken and wrong about everything else, it was he—his thinking self—who was mistaken, and therefore his own existence was absolutely certain. The existence of his body, though, was not certain (he said he could be hallucinating or otherwise deceived). So his body and his thinking self must not be the same thing. Instead, Descartes claimed that we are most fundamentally an immaterial substance that thinks, and we may or may not have a body to go along with that.

This is a tidy solution for explaining the two facets of our existence: our bodies are part of material reality and can be objects of scientific study; our souls are subjects and responsible for thinking. For the Christian tradition of which Descartes was part, souls have the added benefit of mapping pretty well onto the theological conception that distinguishes us from the beasts and being able to survive the death of the body.

The immediate problem for Descartes's dualistic view was that he had no coherent account of how bodies and souls could interact. Over the next centuries, the workings of the mind came increasingly to submit to scientific description and explanation. That made more and more people wonder whether we really needed a separate substance to explain those aspects of our existence that seem to transcend what material particles are capable of. Surely, they thought, the scientific revolution would continue and show how the brain is really responsible for all of these things on its own. And the soul became vestigial—at least that's the company line from most science types. The question we need to explore next, though, is whether this materialist response to Descartes's dualism can only explain the subject-object dualism of our experience by ignoring one of those aspects.

I used to team-teach a course to college freshmen that included Introduction to Psychology. In the textbook, psychology was introduced as follows:

> The word psychology comes from the roots psyche, which means "mind," and logos, meaning "knowledge or study." However, the "mind" is notoriously difficult to observe. That's why psychology is defined as the scientific study of human and animal behavior.[2]

I have no qualms with an academic discipline defining its terms and pursuing the questions it finds interesting. But for the history of human inquiry about the *psyche*—which is more literally translated as "soul"—it feels like the scientific establishment has said, "Yeah, we don't know what that is, so let's just talk about behavior." That's like the classic joke about the drunk guy looking for his keys under a lamppost at night. Someone asks him, "Did you lose your keys here?" And he replies, "No, I think I lost them over there somewhere, but the light's better here. "

Science has shed light on a lot of things, but if we think its light is the only way to know anything, we're making some pretty big and unfounded claims about reality. Science became so successful by narrowing its scope. Its success at explaining that narrower slice of reality led some people to think that only what science can explain is real. Those other things—the "transcendent" qualities of our experience—are just figments of our imagination, like fairies and Zeus.

For example, Francis Crick, who along with James Watson discovered the double-helix structure of the DNA molecule, proclaimed, "You, your joys and your sorrows, your memories and ambitions, your sense of personal identity and free will, are in fact no more than the behavior of a vast assembly of nerve cells and their attendant molecules."[3]

Notice there are two different categories of things in that quotation. One set is amenable to scientific treatment, like nerve cells and molecules; and the others are transcendent concepts, like you, joys and sorrows, ambitions, and free will. These two categories point to the dual reality of humans. Crick explains away the dual aspect by simply asserting that the terms in this latter set are unreal.

They are just silly names we have given to things we didn't really understand, and now we can see that they are nothing but the things more accurately described by scientific language. That is scientific reductionism, and I find it deeply unsatisfying because it is incapable of doing justice to our experience as subjects.

British philosopher Mary Midgley confronts this kind of reductionism in her book *Are You an Illusion?* She is not trying to justify religion or life after death; instead, she claims that other words associated with the transcendent aspect of our existence can't just be eliminated: free will, love, intentions, and so on. Above all, our sense of self cannot be explained by reductive materialism. She adopts Descartes's strategy: if we are deluded into thinking there is a self, just what is it that is being deluded? Can a pile of atoms be deluded?

So we have a really good scientific story for how bodies evolved, but science hasn't yet figured out how to explain the transcendent aspects of our existence. Should that push us toward thinking Descartes was right and that there is an immaterial substance that is the real me? Could we say God creates a soul for each individual and inserts it into each human body at conception or birth (or some other point in between)? Might that explain the two perspectives on humans?

Many Christians believe that, but it keeps us from having an integrated view of ourselves. I think it's worth thinking through some other possibilities. Recall from part III that God doesn't create things initially the way they are ultimately intended to be. What if this applies to the soul as well? What if this capacity of ours to be subjects emerges over time—in each of us individually, as well as in us as a species?

That story starts with the material in our bodies that lasts the longest and seems least like a soul: bones.

BONES AND RELICS

So much of what we know about our ancient ancestors comes from bones. These are the parts that last the longest after we die, and it's remarkable how much information can be gleaned from even a single bone fragment or tooth. A fossilized leg bone can tell us whether the creature spent more time in trees or on the ground and whether they walked upright on two legs. Traces of food preserved under plaque on teeth tell us about the creature's diet.

Can bones tell us anything more? Do our skeletons contribute much to the kind of creature we are? Our skeleton is very similar to the skeletons of other primates. Mammals as a group generally have the same basic skeletal structure, even though some are adapted for running on four legs, some on two legs, some for swimming in the water, and some even for flying. Consider the skeleton of a bat.[1] The finger bones are elongated to support the structure of the wings, and the skull has a different shape. But the rest of the skeleton looks very similar to ours.

Encompassing even more distant cousins in our family tree, tetrapods include the four-limbed animals like mammals, reptiles, amphibians, and birds. The tetrapod body plan has a head at the end of a backbone and four limbs, just like ours.

In my quest to understand evolutionary science, I was beginning to see how similar bodies like these could evolve over long stretches of time with small adaptations. But could these slightly different body plans have much to do with the very different kinds of life and experience that these animals have? What must it be like to experience the world primarily through the sense of smell like a dog? Even further afield from our experience, one of the most famous philosophy papers of the twentieth century asks, "What is it like to be a bat?"[2] The answer is, basically, that we have no idea because a bat's primary way of sensing the world through echolocation is so different from our own. How could that way of life develop with only minor tweaks to the skeleton and body plan? How could our way of life with its rich transcendent qualities and moral responsibility evolve from the same ancient ancestors? To move toward answering these scientific questions, I'll start with . . . an ancient theologian.

Saint Gregory of Nyssa lived more than a millennium and a half ago (he died in 395) in what today is central Turkey. He was a bishop of the church in Nyssa, a small town in the region of Cappadocia. Along with his brother, Basil the Great, and their friend Gregory of Nazianzus, they formed the triumvirate known to church history buffs as the Cappadocian Fathers. They were influential in developing what is now the orthodox Christian understanding of Christ and the Trinity: Christ is one person with two natures; the Trinity is three persons with one nature.

Because he is a saint, his bones are esteemed as holy relics. I had an upcoming trip to Rome and wondered if there was a crypt there where I might be able to see his bones. But a quick internet search revealed that the relics of Saint Gregory were transferred from the Vatican to a Greek Orthodox church in San Diego in 2000. Bummer. I had just been in San Diego a couple of months earlier. I wish I had known. That would have been an interesting side trip.

Instead, I emailed the church and asked for a video call with someone to talk about Gregory's relics. A few days later, Father Simeon responded that he would be happy to talk with me. We scheduled the call for the next day.

About a minute into the call, while I was still introducing myself and explaining what I was working on, he interrupted and said, "Sorry, you need to see this." He turned his phone around so I could see a man in the church sanctuary kneeling down in front of a gold case. "That's where the relic of Saint Gregory is," he told me. He walked over closer so I could see as they opened the case, which had a little glass capsule in it. The capsule contained a fragment of bone about the size of a pencil eraser.

"What bone is this?" I asked, wondering if this was all there was and whether archaeologists would be able to deduce anything from it about the kind of person Gregory was.

"Yeah, we don't know," he said. "It's just a little bone chip of some kind." That was disappointing. I was hoping for at least a finger or a tooth that might have preserved more information. But I suppose this little chip of material must mean something to them.

"So how do you understand the significance of this little piece of bone, and how is it used in your worship or practice of spirituality?" I asked, trying not to sound too skeptical.

Father Simeon replied, clearly having answered this question before from the likes of me, "We in the Eastern tradition of Christianity revere the body. We don't think there is something magical about this relic, but there is something spiritually important about being in the presence of the body of one who was holy and in close communion with God."

I wondered if whatever spiritual advantage there was to be gained by being in the presence of this little piece of Saint Gregory's skeleton was available to someone through a video call. I wasn't really feeling anything. But my skepticism is kind of hypocritical, given that I had been hoping to go and see the relics in person. Why? I'm not sure it would have contributed to my faith or experience of worship, but I obviously valued being there, just as I had valued being in the caves where ancient people drew pictures. Something about knowing what had happened in a place or with an object contributes to the in-person experience. But my next question ramped up the skepticism.

"Can I ask how you know this is really a piece of bone from Saint Gregory?"

Father Simeon chuckled. "You must have heard the jokes about that," he said.

"Tell me anyway," I encouraged him.

"We like to say that if you go around to all the churches in the world and collect all the pieces of the true cross of Christ that are held by them, you'd have enough wood to build the ark." He laughed a hearty laugh, then added, "We take it on faith, . . . and we have a letter from the pope validating it. There was a monsignor from the Vatican who brought it to us with all of the proper paperwork, but it doesn't appear that Rome knows how they came into possession of it."

The stories about holy relics are interesting, but I doubt there is a very high rate of authenticity. I've walked sections of the Camino de Santiago in Spain a couple of times. Since the ninth century, pilgrims have traveled hundreds of miles along the route to see the supposed bones of Saint James the Apostle in the cathedral in Santiago. There is a long tradition of pilgrims experiencing miraculous things along this path. That might make you think there is something miraculous about the bones themselves. The problem is that they probably aren't really the bones of Saint James. The earliest mentions of Saint James traveling to Spain are probably based on a scribal error: someone wrote "Hispaniam" (Hispania) instead of "Hierosolyman" (Jerusalem) in a list of territories for the apostles.[3] It's very unlikely that James the Apostle was ever in Spain, and even less likely that the bones in the crypt of the cathedral in Santiago are his. Yet millions of people keep hiking the Camino, and many keep reporting miraculous things happening.

That means that believing there is something special about the bones or about the pilgrimage is enough to make special things happen. That isn't too surprising, as numerous studies have shown that what we believe has a significant impact on our experience. For example, when test subjects believe a drink has caffeine in it, they report feeling more awake after drinking it.[4] It's fairly easy to manipulate our bodily experience by how we've been primed to think about it.

In the New Testament book of Acts, people believed that if even the shadow of Saint Peter fell on their sick loved ones, they might be healed (Acts 5:15–16). Is that enough to actually heal a physical malady? It seems that sometimes it is.

Undoubtedly, we build up ideas like this in our minds, and there is some actual effect. Beliefs and expectations can influence our bodily experiences. That direction of influence is very well established. My question is whether the opposite direction of influence is also true: do bodies influence the kinds of mental experiences we can have?

Of course the purely material influences souls at one level: if I drink too much Scotch, it affects the way I think, at least temporarily. And forcing yourself to smile can actually make you happier.[5] But I'm looking for something deeper and more enduring: does our body type influence the *kind* of mind we have? Back to Saint Gregory.

Gregory had a remarkable understanding of the relationship of the body to spirituality. In the Western traditions of Christianity, that was largely lost due to the influence of Descartes, as described in the previous chapter. Descartes said the real you is an immaterial substance of some kind. The body might be a house (or a prison) for the soul, but it doesn't contribute significantly to the kind of creature we are or how we think. That is all due to the immaterial stuff.

Gregory would have had none of that. In a treatise he wrote 1,650 years ago, "On the Making of Man," he recognized that the kind of body we have determines the kind of creature we are.[6]

Why don't we have horns, venom, or even hair by which to defend and protect ourselves? He answered that we wouldn't rule over other creatures if we didn't need to cooperate with them to accomplish our tasks (VII.2). If we hadn't been so slow, we wouldn't have needed to tame the horse; if we had a thick coat of fur, we wouldn't have needed to domesticate sheep for their wool. We can't eat grass ourselves, so we domesticated cattle, who eat the grass, and we eat them.

His era of Christian theology was shot through with Platonism and its denigration of the body, so it is a bit surprising that we find so strong an affirmation of embodiment in his writings. In his view, bodies were not just prisons for the soul, and he didn't claim that our higher mental and spiritual capacities exist independently of our body types. Instead, they are functions of each other.

Gregory argued that since God intended us to be rational animals, we needed a body type that works for that. He argued that if we had been on all fours, that would have affected the shape of our heads—just look at the heads of other animals. He even claimed (mostly correctly) that such heads would have made speech impossible (IX). And if we had walked on all fours, instead of being bipeds, we would not have had hands for writing, which has become so important for expressing and articulating thoughts.

He considered arguments that the mind is centered in the heart or liver or brain, but he concluded that it is not localized and must be in equal contact with all the parts of the body (XII). That is a surprisingly modern-sounding insight.

To be fair, not everything Gregory said was quite as insightful. He claimed the four-footed beasts were created this way so that they stoop beneath us, bowing their bodies downward, whereas our upright bodies show our dignity (VIII.1). What would he have thought of the *T. rex*?! And he thought yawning is for the purpose of letting out the vapors that collect in our inmost parts when we don't allow our bodies to sleep (XIII).

But reading Gregory's insightful bits was enough theological motivation for me to keep looking for the influence our body types can have on our souls, now from a scientific perspective. It was time to grapple with bones much older than Saint Gregory's.

18

LEARNING TO WALK

For most of my lifetime, the poster child for ancient human ancestors has been Lucy, found in 1974 in Ethiopia's northern region of Afar. The night of her discovery, the Beatles' song "Lucy in the Sky with Diamonds" was playing in camp, and the nickname stuck. More formally, her bones are known as AL 288–1, which means the first fossil found at the 288th site of the Afar locality. In the local Amharic language, she is called Dinkinesh, which translates as "you are wonderful."

The wonderful thing about Lucy is the picture she paints of our ancient past. Her skeleton is about 40 percent complete, which gives scientists considerably more to work with than many other ancient fossil discoveries. There is enough of her skull to determine that her brain volume was about the size of an orange—slightly larger than a chimpanzee brain today. The ends of her arm and leg bones show that her growth plates had fused, so she was done growing. But she was only between three and a half and four feet tall, which is one of the reasons she is believed to be female. Her teeth show that she was probably in her late teens when she died. And although the cause of death is unknown, there are two bite marks on her pelvis, suggesting that a scavenger took hold of her before she was dropped and covered by mud on the shores of a lake.

When did she live? Fossils more than 50,000 years old cannot be dated by sampling carbon in the bones themselves. Dating therefore has to be done by sampling the material around where the bones were found. Volcanic ash is great for this because lava contains radioactive potassium isotopes that decay into argon at a constant rate (and much more slowly than carbon isotopes). So samples can be dated with a high degree of accuracy. Lucy's bones were found just above a layer of volcanic ash that has been dated to 3.2 million years ago; that is a good estimate of the upper limit of when she lived.

How did Lucy walk? For some time after her discovery, the answer to that question was unclear. In the 1970s, there were different schools of thought on when our ancestors became bipeds. Many scientists surmised that we had to get bigger brains first, which then drove the other changes that are typical of humans today, like upright walking and tool use. But further analysis of Lucy changed that.

You can tell from a skull whether the creature is a biped by the angle of the hole where the spinal cord enters. But sadly, that part of Lucy's skull was not preserved. The knee is also a telltale sign, but the remains found of her left knee were crushed and not easily reconstructed. The pelvic bone was also crushed, and its reconstruction to look like a modern woman's was not without controversy. Was there an arch in her foot that would indicate bipedalism? We can't tell from the bones that were preserved. But there were a couple of toe bones, and they had some very humanlike characteristics, like being tilted upward for pushing off the ground while walking. The toes were longer, though, and curved like an ape's. It's as if she was . . . an "in-between" creature.

Some experts argued that Lucy walked on two legs, but others disagreed at first. There just weren't enough bones to determine this definitively. But the next year, another 216 bones from at least thirteen different individuals were found in Ethiopia and dated to about the same time period. These were dubbed the "First Family," and they gave important comparative data. Someone could always claim that one individual is an outlier or is unrepresentative, but when you have a whole group displaying the same thing, it is more difficult to deny. Luckily, part of the skull from one of the First Family fossils does show where the spinal cord entered the brain case, and this area more closely resembles our skull than that of chimpanzees or other apes that walk on all fours.

Eventually, Lucy and the First Family were recognized to be the species *Australopithecus afarensis*. The genus name *Australopithecus* means "southern ape" and was originally coined for a different species first found in South Africa. The species name *afarensis* refers to the Afar region in Ethiopia where Lucy and the First Family were found.

Determining whether fossils belong to the same species, or

whether they are a previously unknown species, takes some time—
and quite a bit of arguing. It is way more exciting to announce that
you've found a new species with unique characteristics than to announce that you've found another example of something already
known. And like all other people, scientists have egos. But scientists
eventually sort this kind of argument out—usually when more data
becomes available. The new data that determined whether Lucy
and her kind walked on two feet sounds like clues from a crime
novel: footprints.

About 2,500 kilometers southwest of the site where Lucy was
found in Ethiopia is a rich fossil site in Tanzania called Laetoli.
In July 1976, a visiting team of scholars walked around the site
in Laetoli. One of them got bored and started throwing elephant
dung at the rest of the group. Two others ran for cover in a gully
and began to search for ammo of their own. Scanning the ground
for elephant patties, they saw a little patch of hardened volcanic
ash peeking through the rest of the ground. Looking closer, they
saw large animal footprints preserved in it. Shouts brought the
rest of the group down into the gully to inspect the artifacts their
silly game had uncovered.

I wonder how many such scenes have played themselves out to
an ending of false alarm. "Nevermind. Nothing much to see here."
For a while, that's the direction it looked like things were heading
in Laetoli. It's pretty cool, to be sure, to find ancient footprints of
any sort. But for a couple of months of excavating in Laetoli, all
they uncovered were thousands of footprints made by ancient creatures as small as millipedes and as large as elephants. Most came

from midsize creatures like rabbits and small antelope walking on four legs.

But then in September, several prints were found that clearly came from a two-legged creature. Some initially thought it may have been an ancient bear—again, pretty cool, but bear prints would not give much insight into our human past or make any headlines. But the work continued diligently, and after a couple of years, other biped footprints were found. With careful analysis, these were shown to be consistent with the bone structure of Lucy and the First Family. Eventually, fifty-four footprints were found that clearly show a humanlike heel strike, a straight big toe, and even the beginnings of an arch. The clear consensus of experts now is that these came from the same species as Lucy and the First Family: *Australopithecus afarensis*. The Laetoli footprints give a remarkable snapshot into a day in the life of our very ancient ancestors.

Here's what we know with a reasonable degree of confidence: There was a volcanic eruption in the region 3.66 million years ago—some 400,000 years before Lucy lived. It covered the ground with ash, and then shortly thereafter rain turned the ash layer to mud (raindrop impressions are still visible). A bunch of creatures walked through the mud, leaving their tracks, and then the sun hardened them. Another volcanic eruption covered the hardened tracks and preserved them for us today.

Among the many creatures scampering and lumbering through the mud that day were two *Australopithecus afarensis* walking side by side. Judging from the size of the footprints, which correlates pretty well with height, the one on the left was about the size of Lucy; the one on the right was a bit taller—close to five feet. It is likely that these were a female and a male, as there seems to be a significant size difference between the sexes of *Australopithecus afarensis.* Maybe they were holding hands . . . but now we're speculating, of course, and perhaps engaging in unwarranted anthropomorphizing. But their steps were synced like those of a couple walking and holding hands.

A few more facts: From the depth and spacing of the footprints, we can conclude that they were walking fairly slowly and at one point stopped and turned—the tracks get deeper and jumbled together in one place. Then they started going again. What did they turn to look at?

Inside the bigger set of footprints on the right is another set of smaller footprints—a third biped walking with the others. Did Mom and Dad stop and look back over their shoulders to see that Junior was coming along, stepping in his dad's footprints the way a human child might? That paints a touching scene—a loving family out for an afternoon stroll—that can't really be contradicted by the evidence. But it certainly goes beyond the evidence we have.

If we're going to add some color to the scene, we should remember that there was a significant volcanic eruption, probably just the day before. What would creatures like these make of such an event? They weren't running away or even walking quickly, perhaps because Mom had a bit of a limp, which can be deduced from the angle her foot hit the ground. Was she injured in the chaos that ensued when the sky turned dark and the earth shook?

Whatever their immediate response to the volcano, by the time we catch up with them they seem to be walking deliberately in a particular direction. Were they headed to meet up with others from their clan? Were they searching for a water source that hadn't been polluted by all that ash? What could this experience of the volcano have meant to their orange-size brains?

The question we're grappling with here is just how much they thought like we do. It's tempting to go to one of two extremes in trying to imagine how *Australopithecus afarensis* may have reacted to their world. On the one hand, we can't help thinking about how we ourselves would have reacted in similar circumstances. At first, we'd be frightened, and we'd look for comfort and security in those we love; then the adrenaline would kick in, and we'd go into crisis-management mode and make a plan to respond to the situation—even if that meant radically reordering our lives. At the other extreme, we might think of *Australopithecus afarensis* as brutish animals without any capacity for complex emotion or forethought; they had no souls and simply reacted the way a deer does to a forest fire.

The reality is undoubtedly somewhere between these extremes. As already noted, they weren't simply fleeing like a deer running from a fire. But were they acting deliberately after forming a plan? Were they holding hands and looking for solace with their loved ones? I kind of doubt it. But 3 million years ago, their species was on the way toward such behavior, and it seems like learning to walk on two legs had something to do with that. Perhaps in the scene we've recreated from the footsteps, we're seeing a snapshot of the evolution of the soul.

19

WHAT WALKING DOES
FOR THE SOUL

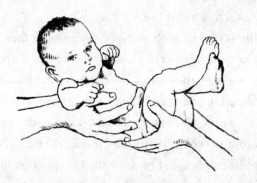

Today, most of us don't think of running or walking primarily as a means of locomotion—of getting from one place to another, a means to an end. We have machines (the complex tools we've built) to transport us with much greater speed and efficiency. If we decide to go for a jog or a walk, we're usually interested in something more than getting ourselves to a different place.

That's what happened to me this morning. I was awake earlier than usual and wrote for a bit before the sun came up. As it got light outside, I noticed through my office window that there was snow on the ground, and it was still coming down. While snow isn't unheard of in mid-November in northern Indiana, it's always fun to see the first snow that sticks to the ground. It wasn't a crazy snowstorm, just big wet flakes gently floating down from the sky. I felt like the scene was issuing an invitation. I had nowhere to go, but I wanted to be part of the story. So I put on my fleece-lined hiking pants and waterproof boots and went out for a walk.

Walking helps me think about things. It's different from the kind of thinking I do sitting here at the computer. Here the thinking is work: I'm actively trying to connect ideas or solve problems. Out there on a walk, it's like the thinking happens on a side channel. You have to pay some attention to what you're doing so you don't get hit by a car or step in goose poop. But when you follow the same course, as I usually do, it gives your mind the leisure to spread out a bit. Ideas often sneak up on me unbidden while I'm walking, and they help solve a problem that I couldn't find my way through while sitting at the computer. I'm sure it's similar for people who say they have their best ideas in the shower.

Walking has become something more than the physical act itself. I'm curious about other animals in this respect. For example, all animals eat, but we have become foodies. Lots of animals build shelters, but we developed Gothic architecture. Some animals use tools, but we have created home improvement channels on YouTube. In the same way, lots of animals walk, but we take pilgrimages. Our walking isn't just walking; quite often it means something more.

Where did these transcendent qualities of the soul come from?

How did we develop the ability to imbue the same physical activities other animals perform with an added dimension of meaning? It seems we developed that ability when we started walking on two legs.

A few years ago, I lived at a house that had a wooded ravine off the back deck. We'd see deer walking through there, sometimes just twenty or thirty feet away. One time a big doe stopped and circled around a bit. We wondered if she was in some distress. Then she squatted down in the vegetation, and less than a minute later she stood up . . . and a new baby fawn stood up next to her! They immediately started walking together. Mom stepped over a log that was too big for baby, and she stood there patiently until the fawn figured out how to maneuver around it. Then they both walked off into the denser woods. It was a remarkable moment to observe.

Deer mothers don't have to contend with big-brained babies when giving birth, so baby deer are born pretty far along the developmental path—they can stand up and walk almost immediately. By contrast, the births of human babies have been pushed earlier and earlier in the developmental cycle so their big heads can make it through the mother's birth canal. Even then it is often with considerable difficulty. Because we're born so early in our overall development, we're pretty helpless at the beginning. Still, by about one year old, our children can start toddling around on two legs. This seems so commonplace to us that we've lost sight of the incredible evolutionary adaptation it was.

Before the discoveries of Lucy, the First Family, and the Laetoli footprints, the prevailing wisdom of anthropologists was that big brains came before bipedalism in the development of our ancestors. The other fossils from the genus *Australopithecus* that had been found prior to the 1970s were thought to have shown only the capacity for intermittent bipedalism, similar to today's chimps. And even after the discoveries of Lucy and the First Family, there was no consensus that *Australopithecus afarensis* were fully bipedal. When Lucy was portrayed in the pages of *National Geographic* in the 1980s, she was depicted as mostly living in trees. But after painstaking work over several decades on the fossil skeletons and the Laetoli footprints, it is now accepted that these ancient ancestors of ours were already walking upright more than 3 million years ago.

That means it was not bigger brains that caused us to start experimenting with different body types that might give a survival advantage. Just the opposite, in fact. It was a different body type that allowed the development of bigger brains. Why would those creatures millions of years ago start spending more time on two legs? It's not like they understood that if they did that their descendants would have bigger brains! And at first blush it seems like there would be a survival cost to living primarily upright with the skeletons we inherited.

For example, running on two legs is much slower than running on four legs. The fastest human sprinters top out at less than 28 miles per hour for just a few seconds. In contrast, four-legged predators on the savanna, like lions and leopards, can hit 55 miles per hour. And now their typical prey, like zebras and antelope, can do about the same. When evolutionary competition takes the path of speed alone, four legs is substantially better than two.

Even today—millions of years later—our skeletons are still not fully adapted to being upright. Our spines have adjusted quite a bit to carrying the load of our bodies, but our vertebrae and the discs between them bear a lot of pressure, resulting in the lower back pain that eight out of ten people will experience at some point in their lives.

There must have been advantages to standing upright that outweighed the cost.

Upright walking slowed down our top speed but turned us into long-distance endurance specialists. Changes to the skeleton—not least of which was the arch in our foot—facilitated this in part. But we also needed a way to keep those bodies cool while they ran in the African sun. Being upright helps with that. There is less surface area for the sun to directly hit when you're upright compared to when you're down on all fours. But the really significant developments were the drastic reduction in body hair and the proliferation of sweat glands that would have accompanied the most successful bipeds in their new environment on the savanna. Only a few other animals have sweat glands all over their bodies like we do. Horses, for example, sweat a lot when they run for an extended period of time. But because they also have thick hair all over, the sweat doesn't evaporate quickly and lower the body temperature.

Sweating allowed us to run longer, but it also allowed for the development of bigger brains. Brains are enormous energy users, and just like the CPU on a computer, they have to be kept cool. Bigger brains couldn't have developed without this cooling system already being in place—they wouldn't have worked for long! So in a very direct way, standing upright on two legs brought us bigger brains and changed the way we think.

Bipedalism was also a huge boon to our ancestors' tool use.

Chimps can manipulate a stick to get termites out of a mound, but we build computers and rockets. How did we get set on the path of extreme tool use that affects almost every area of our lives? The dexterity of our hands has to get much of the credit for this. We can touch each finger with our thumb, which allows us to manipulate tools much more precisely. But you don't get hands like that unless they've been freed from the ground. We might not have been able to outrun big predator cats, but we could make a spear and hold it in our hands while we ran. That was enough for us to compete.

But it's not like we had all these great ideas in our minds for tools we could make if only we had hands. Somehow having fully dedicated hands is what spurred us to begin thinking that way in the first place. In my library are several books about this, including *The Hand: A Philosophical Inquiry into Human Being* and *Prehension: The Hand and the Emergence of Humanity*. The basic argument is that walking upright freed us to use our hands, and the use of our hands led to a sense of agency and self. The opposability of our fingers—which is so much greater than the bodily appendages of any other animal—may have been key in this, as it allows the hand to communicate with itself. And then there is a dialectic or ratcheting up between uses of the hand and the mind's ability to conceive other possible uses. Our sense of self surely developed with this process.

Bipedalism also may have given rise to another distinctively human capacity: cooperation. Like the rest of our abilities, there are other species that can cooperate too to some extent. Cooperation didn't spring from nowhere. But just like tool use, we cooperate to an extent that is unheard of even among the most social species. And it seems that our transition to being full-time bipeds was a significant factor in this. We have to do some speculating here because not a lot of data has survived the millions of years since our ancestors started experimenting with bipedalism. But by comparing what we observe about parenting in our closest living relatives, the chimpanzees, we can draw some reasonable conclusions.

Groups of chimps might forage in the trees, but each grabs its own food; it doesn't take more than one individual to pick a piece of fruit. But it's a different story when looking for food while walking upright on the savanna. You're not going to bring down a mammoth on your own!

Or consider child-rearing. Chimpanzee babies ride on their mother's back, keeping themselves there by holding on to their mom's fur. Since mom walks on all fours, her back is a relatively stable platform for riding. What happens if mom is standing up on two legs? Baby can't hold on by itself for very long, and our ancestors started having much less body hair around the same time they started walking on two legs. So how does mom keep her baby from falling? She has to use at least one arm to hold the baby. That makes mom a lot more vulnerable to predators and less efficient at gathering food. And our ancient bipedal ancestors still lived in trees to a significant degree at first; how easy is it to climb up or down a tree while using one arm to carry a baby? Not very. So here is another significant cost to the bipedal adaptation. How did early bipedal moms and babies survive?

Almost certainly their survival required cooperation—handing off the baby to someone else. This creates a radically different way of life. Chimpanzee babies are hardly touched by any other member of the tribe for six months after birth. For humans today, raising a baby is a collective affair from the very first moments. Grandmothers, for example, typically have a much greater role in child-rearing in humans than in any other species. There are stories of moms who deliver and raise a baby completely on their own, and these are heroic. But as a general rule, this is not a very promising survival strategy. Among our ancestors, those who cooperated with each other would have been much more successful. Passing the baby to a relative so mom could climb the tree, or watching the baby so mom could go out and forage—such cooperation would increase the likelihood of survival for both. It also led to a very different kind of community.

The kinds of behavioral innovations that walking upright helped develop in our ancestors are not just tacked onto otherwise similar organisms. They reordered the kind of life ancient hominins had. Walking upright put us on the path toward the transcendent qualities of the soul. But we're not quite there. Walking upright led to bigger brains, to hands that were tools that could make other tools, and to cooperation. Collectively, these advances led to something else that was the real game changer: language.

FINDING DEEPER FAITH
WITH HELEN KELLER

I've always been fascinated with language. Both spoken and written language are conveyed to us through bits of the physical world that are manipulated in various ways: patterns of sound waves through air or splotches of ink on a page (or pixels on a screen). Yet those marks convey something that transcends their physical reality.

In one of my favorite novels—*Speaker for the Dead*, the second book in the Ender's Game series by Orson Scott Card—there is a scene that always moves me to tears, even though I know it's coming. Somehow, by seeing a certain pattern of physical marks on a page, I'm transported into an emotional space that has a profound effect.

For written words to have that kind of effect, though, you have to know what those patterns of physical reality mean. Words only convey something when they come from someone in your language community. If someone who doesn't know English were to look at that same page in the book, they would not be moved the same way, even though their visual experience is exactly the same. That points to something really extraordinary about language and other symbols: their causal effects do not depend on the specific properties of the physical reality through which they are conveyed, but rather on what a community of people has taken them to mean.

In this way, we might even say that we live in a different world because of language.[1] Yes, we are still in the world of objects, but we spend most of our time in the world of symbols. The physical reality many of us encounter almost constantly is an illuminated screen. How often do you pay attention to the physical reality of that? I'm not even sure if most of us can. I just spent the last five minutes with my nose almost pressed against the screen, straining my eyes to see the individual pixels of the computer screen I'm writing on. As I type these words, I'm manipulating physical reality, causing changes so it looks the way I want it to. But I don't really know how that works. I'm operating in the realm of meaning, not thinking at all about the physical reality that somehow changes as the words move from my mind onto the screen.

The remarkableness of that isn't due to computer technology. The same is true of making just the right sounds come out of our mouths when talking to our neighbors, signing things correctly with our fingers in the d/Deaf community, or making hieroglyphs in wet clay in ancient Egypt. In our symbolic world, we normally don't think about the mechanics of manipulating physical reality, so its ability to convey meaning doesn't surprise us the way it ought to. But watch it through the eyes of a young child just going through the phase of language acquisition, and we can get a glimpse of how strange and wonderful symbols can be.

Recently, my wife and I sat with our precocious two-year-old grandson on the front step, picking up rocks from the flowerbed and arranging them into the shapes of letters on the sidewalk. A light would go on in his eyes when he recognized the rough outline of his favorite letters, and he'd point and exclaim, "M for Mommy!" or "D for Daddy!" He was amazed by the transformation these rocks could undergo. One moment they were just part of the flowerbed, only contributing to the natural setting there; the next moment they became symbols that somehow conjured up a reality beyond themselves. Eager to see the magic happen again, he brought another rock from the flowerbed and manipulated the sound waves well enough for us to understand him saying, "Do this one next."

The transcendent realities are not confined to letters or even words. We live in a world that is populated with things like nations and political parties; sitcoms and baseball games; religion and retirement accounts. For most of these, it doesn't even make sense to talk about their physical reality. They have grown out of words and ideas and have continued to exist because they have a social reality (rather than a physical reality) within a community of people who agree on rules that govern them. It is language that ushers us into that community. Even more so, it is language that gives us a soul.

That last sentence might seem a little surprising and unsubstantiated. Let me work up to it again, beginning with what it's like not to have language.

It is so hard to describe in language what it's like not to have language. The British author Charles Foster tried to live that way and then write about it, but he admitted that it wasn't really the genuine article—he was a fully linguistic being trying to pretend he wasn't.[2]

If we were attempting to investigate through scientific methodology without any concern for ethics what it's like for humans not to have language, we might experiment by raising some kids in an environment where they have no exposure to language. Thankfully we do have some concern for ethics; such experiments should never be tried. But there are a handful of documented cases in which children somehow grew up apart from any meaningful human contact. These let us make some conjectures.

"Raised by wolves" has become a trope for unruly children, but it's based on an actual event. In the 1920s in India, a missionary was summoned to a village because the residents believed there were ghosts living nearby in the jungle. He took a group and found a giant anthill with caves in it, out of which came several grown wolves, two cubs, and two small humans—running on all fours with a wild and inhuman look on their faces. He captured the children and took them back to his orphanage. Both of them were girls (who were often discarded in that culture) who must have lived this way for years. One of the girls died several weeks later, but the other lived for eight years in the orphanage, adopting only the rudiments of human culture and never learning more than a couple of words of language. A man who visited the girl after she had been at the orphanage for several years reported that she didn't display any human graces or virtues, but neither did she show any of our vices.[3] Do virtues and vices require language?

Another feral child who somehow survived in the woods by himself for years was discovered in England in the seventeenth century. He was estimated to be between twelve and fifteen years old, and he was paraded around the salons like a circus freak. The authors Jonathan Swift and Daniel Defoe both wrote about him, the latter calling him "a body without a soul."[4] He came to be known as Peter the Wild Boy and was eventually taken under the care of various people; he died at the ripe old age of about seventy, never learning to speak. Defoe said that without language, Peter lived in a limited world, existing without knowing that he does, looking out at a world that is a senseless theater, seeing only the surface of things.[5]

What is it like not to have language? You only see the surface of things, the physical reality, not the social reality of what it all means. Language allows us to see physical reality *as* something else—that's what a symbol does: it calls to mind something very different than itself. But you have to know the language in order for symbolism to work.

A couple of years ago, I went to a baseball game with a Swede, two Brits, and an American who grew up in Hungary. It was really funny trying to explain things like why it's called a "strike" even if you don't swing, or why a foul ball is also a strike, except it's not a strike on the third strike . . . except it is if the catcher catches the foul tip on the third strike (which doesn't count as an out on strikes one and two!). They didn't understand what they were seeing because they didn't know the rules of the game.

I've been on the other side of that experience too, watching an e-sports contest on TV with my kids. They would "ooh" and "ahh" at the action in unison because they understood what they were looking at, while to me it just looked like a blooming buzz of confusion.

In both of these cases, though, we were able to talk about it and gradually make progress toward understanding. Imagine not understanding what you're experiencing and not being able to talk about it. This is the story of Helen Keller, who lost both her sight and hearing at about a year and a half old. She could communicate only through gestures and grunts until she was seven, when her teacher, Anne Sullivan, began teaching her language by finger signing into Helen's hand.

The story of her coming to realize that these signs meant something is remarkable. She had previously only had Pavlovian responses that formed through habit, but then something clicked. She realized the signs were symbols that stood for things—these things had names! That opened up her mind to the world we take for granted. She said in her autobiography,

> Before my teacher came to me, I did not know that I am. I lived in a world that was a no-world. I cannot hope to describe adequately that unconscious, yet conscious time of nothingness. I did not know that I knew aught, or that I lived or acted or desired. I had neither will nor intellect. I was carried along to objects and acts by a certain blind natural impetus. . . . My inner life, then, was a blank without a past, present, or future, without hope or anticipation, without wonder or joy or faith.[6]

If language allows us to see physical realities as something more, something that transcends the physical properties, perhaps that applies also to our own existence. With language, Helen could see herself *as* a person, a self, a soul. Before language, she couldn't see herself that way, and it wasn't because she didn't know the word *soul* or *self*. It was because you have to be a language user to have access to the transcendent world, and that's where souls are.

In his wildly popular book, *Sapiens*, Yuval Noah Harari writes,

> *The truly unique feature of our language is not its ability to transmit information about men and lions. Rather it's the ability to transmit information about things that do not exist. As far as we know, only Sapiens can talk about entire kinds of entities that they have never seen, touched or smelled.*[7]

As examples of things that don't exist, he names religion, corporations, nations, and money. It is curious to me that without argument he gives the criteria of things that are real as those that can be seen, touched, or smelled. Only physical things are real? Why should we think that? Why not think that a different kind of real thing can emerge out of physically real things? I don't see this as being very different from life emerging out of nonliving particles of matter when they are arranged properly.

There is a lot more to be said about language (pun intended!). The question of how it first began in our species—or perhaps before our species—is fascinating but not currently resolvable. It looks as if walking upright on two legs started a cascade of changes that prepared us for language, but we don't know what ultimately flipped the switch for our species the way it did for Helen Keller. We don't even know when that happened. Some think it occurred as recently as 50,000 years ago; others think language developed in our ancestors prior to *Homo sapiens*, hundreds of thousands of years ago.

It's also fascinating to consider whether other animals have any rudimentary language abilities. Of course they communicate—and so do plants. But do they have the symbolic ability to see physical reality "as" something else? It's hard to say, because again we're fully linguistic beings trying to think about how they might think about things—we can't help it. Considering those questions would take another book.

Here, the challenge of the soul is to understand how, through the process of evolution, our species could come to have a soul—the rich, first-person awareness that I am a thinking thing capable of living in the symbolic, human world, in addition to living in the natural world. Saint Gregory's insight was that the kind of body we have determines the kind of soul we have. Then we saw that a minor adaptation like walking upright on two legs can lead to some remarkable changes in the kind of life we have. And finally Helen Keller's insight about language showed how souls could result from this evolutionary process instead of simply being tacked onto our bodies. With language we are ushered into the symbolic world, which transcends our physical world.

Just as the meaning of this sentence rides along or emerges from the physical reality of the ink or pixels you're reading it from, so too the soul rides along or emerges from the physical reality of us as persons. And just as for meaning to be conveyed from these symbols to one person, there has to be another person (even a community of people) intending to convey that meaning, so too for us to have real souls, maybe there had to be someone behind the process intending to create us in that way.

I don't mean this in some simplistic way, as though God were one of the causes or forces that science can measure and that if we just look at the science more carefully, we'd find God in the equations. That seems to me like thinking you could find meaning in the rocks themselves that my grandson learned to arrange into letters. But the opposite absurdity is to think that since there is no meaning in rocks or pixels themselves, that there can't be any meaning that emerges from the right collection of them. In the same way, the sacred chain, which links who we are today with the evolution of our ancestors, can't be seen through a microscope. But that doesn't mean there isn't something more that emerges from the process of evolution, some added dimension of meaning.

Just as our eating has become more than calorie consumption and our architecture more than shelter from the elements, I was starting to think that our evolution has become more than survival of the fittest. And similar to C. S. Lewis's "bottom-up" inspiration by which the ordinary products of humans have become inspired and are taken up into divine service, it's defensible to claim that God has also taken up the ordinary products of evolution into the divine service of giving us souls.

That perspective allowed me to see the process of evolution as really powerful and beautiful and faith affirming. To make us with these extraordinary abilities, God didn't have to zap immaterial souls into us completely outside the natural process of creation. But now that brings us to the final challenge I had to confront: why would a good and loving God create us through the process of evolution, when it involves so much pain and suffering?

PART V

PAIN

HUMAN EVIL

In September 1991, a couple of hikers were 10,500 feet up in the Alps near the border of Italy and Austria. They saw what they thought was a piece of trash poking out of the ice just off their path—something carelessly left behind by a previous hiker. But with a little digging, the trash was revealed to be . . . a human head. That suggested carelessness of a different magnitude than littering!

A park ranger was fetched, and they started digging out the head and found it belonged to an intact body. At first they assumed this must have been some recent hiker who misread the weather or got lost and froze to death on the mountain. That sort of thing happens occasionally at such high altitudes. But as they melted the ice to retrieve the body, the preserved clothing and equipment told a different story.

The frozen hiker—who came to be known as Ötzi the Iceman, named after the Ötzal mountain range where he was found—wasn't wearing a North Face jacket and boots. His coat was made from strips of brown and black goat hide, stitched together with animal sinew. His shoes were made of woven tree fibers, stuffed with hay, and covered with deer hide. He had a hat made from bear fur.

Instead of a backpack, he had two birch-bark containers that held burnt wood and tinder fungus for making fire. He had a rudimentary first aid kit with other fungus that had antibiotic properties. And he carried some weapons: in his utility belt were a dagger made from flint, some arrows, and a copper ax with a handle carved from a yew tree.

These are not the accessories of a modern hiker. In fact, through carbon dating, it was determined that Ötzi died 5,300 years ago—long before the time of Christ or even Abraham. When Ötzi walked the Earth, the Egyptian pyramids and Stonehenge had not been built. Nobody spoke Greek or Hebrew or even Tamil or Sanskrit (the oldest languages still used today).

Scientists can tell from his preserved bone cells that Ötzi was forty-six years old when he died. Still in his digestive tract were recent meals of red deer, ibex, legumes, leafy greens, and berries. That sounds like a healthy diet, but Ötzi was not exactly the picture of health. From his DNA and clogged arteries, we know he had heart disease. From his microbiome, we can tell he had Lyme disease, peptic ulcers, and gingivitis. But none of these led to his death, the cause of which is clear . . . and more sinister.

For ten years after his discovery, no one recognized the relevance of a small wound just above Ötzi's shoulder blade. But then an X-ray revealed an arrowhead lodged in his chest! It hit one of the large arteries branching off the aorta, and within about five minutes he bled to death. Ötzi was shot in the back.

He fell face down and must have been rapidly covered with snow. That, combined with the high altitude, essentially freeze-dried and preserved the organic materials, leaving us an impressive window into this man's life and his world, more than five thousand years ago.

We see a clever man who was able to leverage available resources to give himself an advantage. Yes, he's the one who died in whatever conflict occurred up there on the mountain. But at the ripe old age of forty-six back in the Stone Age, it's not a stretch to think Ötzi was an alpha male. He probably dealt more damage than he took over the years. And his medical kit shows he was prepared for lots of different dangers. Through his remains, we might see a world where everything—and everyone—is out to get you.

We ought to wonder, though, just how representative the vista is that we see through this one window onto the ancient past. Was Ötzi a typical man of his age, wandering high up in the mountains, kitted out for hunting and battle? Or are we victims of a sampling error? Perhaps the only kind of specimen that would be preserved for us is one that died suddenly high up in the mountain snow. The lives of normal people down in the valley could have been very different, but they weren't preserved for us. Maybe it's like the fisherman who concluded that a lake only has fish longer than two inches, since that's all he ever caught in his net with two-inch holes.

But it does fit with the general impression most people have of ancient human life that Ötzi could have been some powerful and skilled warlord or assassin. Maybe he had been planning to raid a campsite, thinking he could score some easy meat and possibly even steal some women. Maybe he had done that successfully many times before. But this time he got ambushed by someone else on the lookout for prey. The guiding principle of the precivilized world we're imagining was to kill or be killed.

This all fits with how the seventeenth-century philosopher Thomas Hobbes described the "state of nature"—that time before human beings became civilized. In his book *Leviathan*, he said it was a time of "continual fear, and danger of violent death; and the life of man, solitary, poor, nasty, brutish, and short."[1]

Ötzi is just one data point, but he's been heralded as the poster child or archetype of ancient humanity.[2] This one example conforms to the expected pattern and makes us believe that the natural development of humanity through evolution would only ever produce creatures such as these: just another animal that's interested only in survival and reproduction, no matter the cost to others.

This is a problem for people like me who believe that a loving God created the world and called it good. Does the ancient world that science has uncovered look like the kind of place that was intended by a good God? Or does it look more like the outcome of blind chance and survival of the fittest? Examples like Ötzi have controlled the narrative and set expectations for what we see among early humans. That is part of what I'm calling the challenge of pain. But it gets worse.

In the decades after Darwin published *On the Origin of Species*, his theory was applied to the social world too, resulting in all manner of evils. The dominant social theme of the latter part of the nineteenth century was progress—society would improve as humanity increasingly understood the laws according to which things work. English philosopher Herbert Spencer had been developing his own understanding of evolution, but upon reading Darwin in the 1860s, he coined the phrase "survival of the fittest" and applied this to social policies. He thought the surest way for society to progress was to allow the fittest to dominate. So he advocated for laissez-faire government policies that freed powerful people to use whatever means necessary to advance their priorities. Those unfit for that kind of competition should have to suffer the consequences.

Francis Galton, a cousin of Charles Darwin, claimed nature should inspire the opposite tendencies in a government looking to improve society. Galton is sometimes called the Father of Eugenics, the movement to "improve" the species by regulating which people are allowed to have offspring—which is not Darwin's natural selection at all, but artificial selection. Galton saw that laissez-faire government policies were not leading to the extinction of "lower-quality" people but rather to their proliferation in the slums of cities. To keep them from passing on their qualities—which must be due to their innate inferiority rather than to social structures—he believed the government should limit the number of offspring such people could have. This was deemed to be a scientific way of improving the human race.

Of course, the culmination of eugenics and the ultimate example of evil was Hitler's program to "purify" the Aryan race through the extermination of Jews and other people he deemed impure. Less extreme but still awful examples of eugenics became popular early in the twentieth century as the genetics of inheritance was being discovered. The American Breeders Association sounds like an organization for dog or horse enthusiasts, but it was founded in 1903 to investigate and report on heredity in the human race and to "emphasize the value of superior blood and the menace to society of inferior blood."[3] In 1914 the organization changed its name to the American Genetic Association and published *The Journal of Heredity*. American inventor Alexander Graham Bell served as the organization's honorary president for its Second International Eugenics Congress in 1921 and contributed an article called "How to Improve the Race."[4] These were more than theoretical exercises; they resulted in public policies through which as many as 100,000 people in America were sterilized against their will so they would not pass on their "inferior" genes.

Charles Darwin himself had complicated views about race for his day. He vocally opposed the slave trade when many "respectable" people in his culture did not. Yet he seemed pretty confident that European white people were a superior race to African and indigenous peoples in North America, and would surely replace them.

There is no doubt that some proponents of unrestrained capitalism, eugenics, and racism were inspired by Darwin's work. Evolution-denying groups today often try to leverage this fact as an argument for why others should reject evolution. For example, in an article called "The Dark Side of Evolution," Answers in Genesis says, "Eugenics is simply a logical conclusion from Darwinian evolution."[5] Here's another example from an essay by Henry Morris (one of the authors of *The Genesis Flood*):

> *As the 19th century scientists were converted to evolution, they were thus also convinced of racism. They were certain that the white race was superior to other races, and the reason for this superiority was to be found in Darwinian theory.*[6]

A direct implication of such statements is that accepting the science of evolution should lead you—if you're going to be consistent in your beliefs—to racism, bigotry, and even eugenics. The bigger picture behind such claims, which informs the challenge in this section of the book, is that part and parcel of evolution are attitudes and practices that run contrary to the Christian faith.

Here was another challenge I would need to respond to on my journey of reconciling evolution with my faith. How could I affirm the science of evolution if it is fundamentally at odds with attitudes and practices like sacrificial love and care for the vulnerable, which are supposed to characterize followers of Christ?

The first thing to note is that the science of evolution did not produce racism or eugenics. People held such beliefs long before Darwin. The strong have always tried to take advantage of the weak. We find proposals for government-controlled human reproduction as far back as Plato's *Republic* in the fourth century BCE. And of course the racism associated with slavery was flourishing well before Darwin. The theory of evolution did not lead to these evils, but rather people who were inclined to pursue them looked to Darwin's theory for scientific justification for their heinous deeds.

On the other hand, even if evolution didn't cause these kinds of racial and prejudicial attitudes to emerge, isn't it bad enough that it could encourage more people to have those kinds of attitudes against their fellow human beings? Yes, that would be bad if acceptance of evolution actually led to the encouragement and entrenchment of racism and discrimination. But when a series of empirical studies was carried out in 2022 to look at the connection between these kinds of attitudes and the acceptance of evolution, just the opposite was found. The leaders of the study found that "low belief in human evolution was associated with higher levels of prejudice, racist attitudes, and support for discriminatory behaviors against Lesbian, Gay, Bisexual, Transgender, and Queer (LGBTQ), Blacks, and immigrants."[7] In other words, in today's world at least, it is people who don't accept the science of evolution who are most prone to believe and act in discriminatory ways toward their fellow human beings. Let's not blame evolution for what is more fundamentally a misuse of human freedom—what my theological tradition calls sin.

When people commit evil against other people, there are disastrous consequences. Anyone who suffers at the hands of another human being faces deep existential and emotional pain, and I don't think philosophy and theology can solve this. But at the theoretical level, human-caused evil has not been a serious problem preventing me from continuing to believe in God. I think it is perfectly consistent that, in a world created by a loving God who wanted free and responsible creatures in it, those creatures would sometimes choose poorly and do tremendous harm to others. Whoever shot Ötzi in the back committed a heinous crime and act of evil.

Furthermore, through our choices, systems of evil can develop that become entrenched and perpetrate evil through society. This kind of evil isn't caused by the immediate or direct choices of individuals, but it was still ultimately caused by humans.

Human-caused evil is not what creates the challenge of pain and suffering for me. It would exist whether or not our species evolved. However, there is another kind of pain and suffering more endemic to evolution that is more difficult to explain away.

22

NATURAL EVILS?

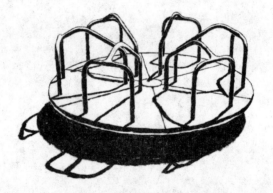

When my son Casey was little, the neighbor's dog bit him on the leg. The spot turned black and blue, and there were little puncture marks where the teeth had broken the skin. We had to do a rabies test to make sure there weren't going to be more serious health consequences. It turned out fine apart from some temporary nervousness Casey had around the neighborhood dogs. Of course, the authorities had to be contacted to ensure that the dog would be properly restrained in the future, but that was it.

How does a dog biting my kid compare to the ways that pain and suffering are caused by people? I said in the previous chapter that I think it is appropriate to call those evil. But is a dog bite evil? We certainly said, "Bad dog!" but the "bad" there doesn't really have moral connotations. Think of how the same event would be evaluated differently if, instead of the neighbor's dog, it was my neighbor who bit Casey! Different authorities would have been contacted, and we would have said more than "Bad neighbor!"

I think it is easier to reconcile the sort of evil caused by morally responsible human agents than it is to reconcile the pain and suffering (whether or not we call it evil) that comes about through natural means. There is a morally responsible person (or an evil system) to blame for the former. It's harder when the pain and suffering comes through no one's fault.

Consider another example. Let's say a house burns to the ground and kills the people living in it. This is an awful tragedy, no matter what. But it seems to me that if the cause of that fire was an arsonist with a grudge against someone in that household, then it is pretty straightforward to say that the arsonist is a morally responsible agent who did an evil thing. That kind of evil is the direct result of our having moral responsibility, and it's at least theoretically defensible to say that the world is a better place for having morally responsible agents even if they sometimes do evil things. That is what philosophers and theologians call a "greater good" theodicy: the world is a better place for having genuinely free and morally responsible people—even if they sometimes cause pain and suffering—than it would be with just a bunch of robots. We might see why a good and loving God would create the world that way.

But how would we think about this tragedy if the house fire was caused by a lightning strike? It is still a tragedy that the house burned down and the family died. And it is an example of pain and suffering in the world. But is it evil in the moral sense? There isn't a morally responsible person we could trace it to . . . or is there?

I've mentioned the possibility that bad systems could be set up through human choices, and then the bad consequences are separated from the choices by a longer causal chain. The bad choices may not even have been made with intent to cause pain and suffering. For example, say a landlord fails to maintain their property and a fire breaks out, killing the family renting the house. The landlord didn't intend for the tragedy to happen, but if the lapses in upkeep directly led to the fire, we might say the landlord was morally responsible and did something evil.

The question before us now is whether there is a morally responsible agent connected to the tragedy of a house fire caused by a lightning strike. We might be able to make the case that, through our decision to burn fossil fuels, we have caused an increase in extreme weather events that have led to more deaths. I suspect that is true, but it doesn't explain all of the pain and suffering caused by natural events. Lightning strikes, hurricanes, and earthquakes have been causing suffering long before we figured out how to burn fossil fuels.

Furthermore, we don't have to limit ourselves to the pain, suffering, and death experienced by humans. The rest of the natural world has plenty of suffering too. Richard Dawkins persuasively and graphically portrays how pain, suffering, and death are not limited to extraordinary times but are a normal part of the world's operation:

> *During the minute it takes me to compose this sentence, thousands of animals are being eaten alive; others are running for their lives, whimpering with fear; others are being slowly devoured from within by rasping parasites; thousands of all kinds are dying of starvation, thirst and disease. It must be so. If there is ever a time of plenty, this very fact will automatically lead to an increase in population until the natural state of starvation and misery is restored.*[1]

Can we reconcile this kind of pain, suffering, and death with a good and loving God by saying a morally responsible agent caused it? The strand of Christian theology I grew up in would say yes: The original sin of Adam and Eve is ultimately responsible for all of the pain, suffering, and death that we find in the world today. They committed a sin by violating what God had commanded, and although they didn't intend the lightning strike that would set the house on fire and kill the family, their morally evil action set into motion a sequence of events that directly led to all the pain in the world.

Before Adam and Eve's sin, these Christians believe, there was nothing bad in the world at all. Then they disobeyed God's commandment and ate a piece of fruit from the forbidden tree in the Garden of Eden, according to a literal reading of Genesis 3, and that act unleashed lightning strikes that burn houses down, and tsunamis that swallow up whole villages. If Adam and Eve's sin is directly responsible for bringing about these "natural evils," there really isn't any distinction between tragedies caused by human actions and tragedies caused by natural means—all bad things are ultimately the result of intentional action by morally responsible people and therefore are evil.

That's a tidy explanation, but there are a couple of really big problems with it. First, we now have plenty of scientific evidence that pain, suffering, and death were part of the natural world long before humans were around to sin. There is clear evidence that dinosaurs had cancerous tumors, for example. And at least five mass extinctions of plants and animals occurred, caused by natural disasters like volcanic eruptions and meteor strikes.

Beyond these extreme examples, it looks as if the kind of situation described by Dawkins is the normal way that life has proceeded on this planet apart from human interaction and interference. Once we accept the abundant evidence of an extremely old planet and the relatively late appearance of *Homo sapiens*, there is no getting around the fact that there was pain, suffering, and death long before we had anything to do with it.

Furthermore, I don't see how the explanation of Adam and Eve eating a piece of forbidden fruit helps us reconcile the existence of these natural evils with a good and loving God. It is fair to wonder whether creation is "very good" if it includes death and suffering. But we also ought to wonder whether a creation is "very good" if it is made to massively deteriorate when two innocent and naive humans disobey. That seems to be a pretty big flaw in the system. Adam and Eve in Genesis 3 don't yet know good and evil. It is only after their act of disobedience that God said, "See, the humans have become like one of us, knowing good and evil" (Gen. 3:22). So to blame all of natural evil on the action of these innocents doesn't seem quite fair.

To try to explain my concerns about this traditional explanation for natural evils, I've developed a parable.

Once there was a construction contractor who built a big, lavish playground for his two kids, Addie and Evan. Every aspect of the playground looked beautiful and perfect. There were swings, slides, teeter-totters, and everything else kids could want. Looking at it from the outside, you'd say, "Wow, that's very good."

The contractor put Addie and Evan into the park and said to them, "I have built this for you to enjoy. Play on all of it—but don't touch the merry-go-round in the middle of the playground. If you do, some very bad things are going to happen."

Evidently, the contractor set up the playground as a test of some sort. Children are naturally curious; would they obey and do what is right? For it to really matter, there needed to be serious consequences. So behind the scenes, he booby-trapped the merry-go-round so that if it spun around even one time, the animal park next door would explode.

The contractor left the kids to play. But for some reason, he also allowed into the playground a creepy, smooth-talking guy dressed as a snake. Addie and Evan played on all the other fun things for a while, but eventually they got to the merry-go-round in the middle of the park. There the creepy guy in the snake costume persuaded these young, innocent children that it really wouldn't be so bad to give the merry-go-round a whirl. In fact, he claimed, trying it out might even make you more mature and become like the contractor himself, knowing what is really right and wrong. The young, innocent kids found this to be a convincing argument, so Addie tried it first, and then convinced Evan that he should hop on too.

As promised, spinning the merry-go-round led to a catastrophic effect at the animal park, which blew up, killing almost all of the animals in it. Who is responsible for this pain, suffering, and death?

Do we say, "Those darn kids! If only they had done what they were told, none of this would have happened!" Well, kind of. But my moral intuitions go a different direction. I find it very hard to call that a "very good" playground and want to demand from the contractor, "Why did you make it that way?!"

I know analogies aren't perfect, but it sure seems to me that the booby-trapped playground is like the creation model according to which God has rigged the very good creation so that as soon as Adam and Eve sample the forbidden fruit, parasites pop into existence, cougars grow fangs, and hurricanes wipe out villages. Do we blame Adam and Eve and human sin for the natural evils? I'm afraid if this is really the way the world was set up, I want to blame God for making it this way. In that case, there is still a morally responsible agent behind the tornadoes and cancers that cause so much suffering and death, but that morally responsible agent is God.

If that's really the way the world was—perfect until Adam and Eve ate a piece of fruit at the suggestion of a talking snake—then I don't see how it gets God off the hook for the natural evils in the world.

The challenge of suffering in an evolutionary world is a difficult one. There's no way around it: God is ultimately responsible for making a world where lots of innocent creatures suffer and die. I haven't tried to answer that challenge yet. I'm only claiming here that those of us who accept evolution are in no worse a predicament than the young Earth creationists in this regard. The pain, suffering, and death through natural causes in the world they describe ultimately trace back to God just as much as they do in an evolutionary world. I don't know that we're ever going to be fully satisfied with an explanation for this, but I do think we have a better option than thinking it is all the result of human sin.

23

WHAT GOD CAN'T DO

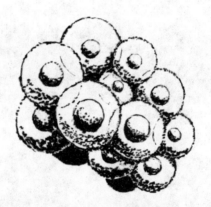

When I was still regularly teaching Introduction to Philosophy, one of my favorite class sessions was the one that covered Saint Anselm's Ontological Argument for God's existence. I'd walk into the room just as class was about to start and announce, "My Aunt Pat has red hair." Students would look up from chatting with each other, wondering whether class was actually starting or whether Professor Stump had gone a little batty. When I had their attention, I'd ask, "What else can you tell me about her?"

Someone would pipe up, usually a guy from the back row thinking he was being funny, and say, "She's female."

"Yes, I kind of gave that away by asking, 'What else can you tell me about *her*.' But with the difference that is acknowledged between sex and gender these days, that's not a given. And it's certainly not a given that aunts have red hair. Can you tell me something else about my Aunt Pat? Something you're absolutely sure must be true of her?" I would continue to look at the guy in the back row, who would now be squirming a bit in his seat, not realizing he had volunteered to engage in some actual philosophical dialogue.

"No. I don't know her."

"That shouldn't stop you from being able to tell me something else. Think harder."

"I don't know, man. Ask someone else."

I would then turn back to the rest of the students and ask, "What has to be true of my Aunt Pat?" Most of the students would look a little confused. But usually one excited student would suddenly realize and blurt out, "Oh, she has a brother or sister!" Then a debate would ensue between students in the class about whether she herself has to have a sibling or whether it could be that she's an only child who married someone who has a sibling.

"Hmm . . . how do we know who's right? Is there someplace we can look to decide?" Someone would already have dictionary.com pulled up on their phone and announce that one definition is "the sister of one's father or mother" and another is "the wife of one's uncle." So it looks like both definitions are okay. Aunts don't have to have red hair, but they do have to have a sibling who has children or be married to someone who has a sibling with children.

So far so good. The class was understanding what it means for something to have a necessary characteristic. Then I'd ask in a more serious and questioning tone, "What can you tell me about God?"

Saint Anselm was a theologian at the end of the eleventh century who developed the Ontological Argument for God's existence. He claimed we can show that God must exist in the same way that aunts have to have a sibling (or be married to someone who has a sibling). His argument rests on articulating two concepts in a particular way: (1) God was defined as the best possible being; and (2) existence was claimed to be good, so that it is better to exist than not to exist. If both of these are correct, then existence is a necessary characteristic of God, because an existing God would be better than an imaginary God.

Most students were suspicious that there was some smoke and mirrors going on with these definitions, but almost every time I taught this, there would be a student or two who saw the Ontological Argument as an epiphany. In defense of the rest of my students, there's plenty of philosophical controversy around the Ontological Argument and whether Anselm's two claims are correct. But the argument doesn't have to ultimately be successful to make my first point in this chapter—there is something God can't do, according to Anselm (forgive the double negative here): God can't not exist.

We can use that same kind of conceptual analysis to show there are other things God can't do. God can't make a square circle, right? That's not a weakness or deficiency in God, but rather a logical contradiction. The terms *square* and *circle* themselves, as we have defined them, rule out the possibility. Also, according to Hebrews 6:18 in the New Testament, God cannot lie. That contradicts the divine nature. So God can't do contradictory things.

In my quest to respond to the challenge of pain and suffering, I've wondered whether the kind of world God wanted couldn't be created without pain and suffering over long ages. I believe that God intended to create morally mature beings like us who could bear the image of God. But maybe not even an all-powerful God could create morally mature beings ready-made, as it were. Instead, our ancestors had to go through challenging times, choosing for themselves how to respond. An all-powerful God can't create a square circle or an unmarried aunt who is an only child because those are gibberish; there is no possible object that corresponds to such a description. I'm wondering whether "a morally mature person who did not participate in their own development" is similarly an oxymoron.

What does it mean to be morally mature? I think it means you have a record of responding to challenging situations with wisdom and insight. You consistently choose what is good. Would God create beings with a fictitious record of responding to challenging situations, like creating a tree that looks like it's a hundred years old by putting a record of fictitious environmental conditions in its rings? I don't think so. Just like the fake tree ring situation I described in Chapter 7, every individual's record would have to be coordinated, resulting in a massive deception. But this is more than that. You can't be given moral maturity. You earn it or grow in it. That means that not even God can create morally mature persons without their own participation. And it means that potentially morally mature people are going to have to be subjected to challenging situations so they have the opportunity to choose wisely.

There are two significant objections to what I've claimed so far. The first is related to my claim that God intentionally created morally mature beings but did so through evolution over a very long time. "Wait," the objector says. "Isn't evolution a random process? Just like square circles, it is a contradiction in terms to say that even God could intentionally create humans through a random process." That is some of the rhetoric that has been used to claim that it is illogical to affirm both the science of evolution and a creator God who might somehow bring about a desired conclusion. If evolution is random, they argue, then it could end up anywhere.

It is not hard to find legitimate evolutionary scientists who have said essentially that. Back in the 1980s, the great American evolutionary biologist Stephen Jay Gould famously argued that if we could rewind the tape of life and play it again a million times, we wouldn't end up with anything like *Homo sapiens* again.[1] But more recently, Simon Conway Morris, a paleobiologist at the University of Cambridge, has become the poster child for convergent evolution. *Convergent evolution* is the phenomenon that the same structures evolve independently from different starting points. For example, wings have evolved independently on birds, bats, and insects; and the eyeball of a giant squid looks and functions like ours, even though our common ancestors had nothing of the sort. Conway Morris and his colleagues have found hundreds and hundreds of these convergences, and he thinks they point to some deeper constraints in how things evolve.

I talked with him on the podcast about his newest book, *From Extraterrestrials to Animal Minds: Six Myths of Evolution*.[2] One of those myths is that evolution is random. What is most amazing about the history of life to him is how few of the possibilities for how life could evolve have actually occurred. That is not because of a lack of time but rather because life is much more highly constrained than previously believed.

He speaks of a "habitable node" as one of the theoretical possibilities for how life could evolve, and he says there could be as many as 10^{250} of these (1 followed by 250 zeros). And yet the actual number of habitable nodes is only about 10,000.[3] Yes, there is an incredible diversity of life on the planet, but there are so, so many more ways that life hasn't evolved.

Even the mass extinctions that were often driven by a "random" event like a mega-volcano or meteor strike did not fundamentally alter the trajectory of evolution. Conway Morris says they "serve to accelerate what was going to happen anyway,"[4] and he has lots of data to justify his claim. There are enormous constraints on what can evolve, and this leads to the same things evolving time and time again—from wings, to eyes, to photosynthesis.

The specifics of these processes are fascinating, but recounting them here would take us too far afield. The general point is that evolution is not random. Responding to Gould, Conway Morris claims, "Contrary to the received wisdom, the emergence of human intelligence is a near-inevitability."[5] He thinks that's because there is a deeper principle at work in evolution that we have not discovered yet—something similar to how the periodic table of elements showed the underlying principles of chemistry. I pushed him further on this point, asking what it might be. His response:

I, to date, have not been able to articulate it in any coherent fash-
ion at all. It is almost an instinct, that underneath what life evolves
along, there is, if you like, an underlying melody for want of a better
word.[6]

It will be interesting to see where this research goes in the next generation.

There is another response to the objection that God intentionally created through the supposedly nonrandom character of evolution, and we don't have to wait for this one to be thoroughly documented in some future research. There is clear directionality in how life has developed with respect to cooperation.

Researchers have done a good deal of work in the last decade on things like cooperation and altruism and how evolution might have produced such capacities. Evolution is too often presented, particularly for lay audiences, as entirely about competition, struggle, and survival of the fittest. What scientists have increasingly found, though, is that often the fittest organisms are the ones that have learned to cooperate with others rather than dominate them.

Scientists are skittish about seeing teleology, or purpose, in the scientific details of the progression of forms, but as Christians who allow theology to bear on our understanding and interpretation of science, we might see something more than the bare facts. But let's start with the facts, which are these: The development of life on Earth has progressed through major transitions, each of which included a substantial increase in the amount of cooperation:

- From simple prokaryotic cells to complex eukaryotic cells with nuclei and organelles, where different components play different roles and have to work together

- From asexual reproduction, where organisms produce clones of themselves, to sexual reproduction that requires a population of individuals and sexual differentiation

- From single cells to multicellular organisms in which different kinds of cells work together to sustain and advance the life of a complex organism

- From solitary individuals of multicellular organisms to colonies of many individuals with different roles working together

- From social primate societies to almost unimaginably more complex human cultures

When we combine the phenomenon of convergent evolution with this clear directionality of increased cooperation, it's hard for me not to see a kind of intentionality there on the part of the Creator. These are the kinds of creatures that God wanted to exist, and it looks like evolution is perfectly capable of bringing them about.

There is another objection, though, to my claim that God intentionally created morally mature beings through the process of evolution: it still seems inconsistent with how God would do things. To begin answering that objection, we need to zoom in closer on that last major transition: from primate societies to human cultures.

BUILDING MORALITY

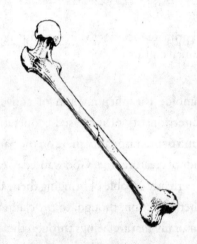

had a really fascinating interview with Sarah Brosnan on our podcast's Uniquely Unique series about the development of morality.[1] Brosnan is Distinguished University Professor of Psychology, Philosophy, and Neuroscience at Georgia State University and also codirector of the Language Research Center at the university. She has done a lot of work with monkeys in trying to understand the evolutionary development of capacities in humans like cooperation and morality. She described an experiment during her graduate research that ended up going viral on YouTube.

Monkeys that knew each other well were trained to make trades with humans. They had a supply of small rocks in the cage, and when they brought one of them to the handler, they would get a treat. For the control group, monkeys got a treat that they liked: a cucumber. That is not their favorite kind of food, but they would always happily eat them . . . until they saw their friends getting a better treat. For the experimental condition, one of the monkeys brought a rock and was given a grape in return. In the hilarious video of this experiment,[2] the control condition monkey observes his friend getting a grape instead of a cucumber, so he immediately goes to get another rock and trade it. When he still gets a cucumber, he gets so upset that he screams and throws the cucumber back.

These monkeys were perfectly happy with a cucumber when their friends were only getting cucumbers too. But if other monkeys received grapes for the same trade, they were very upset. You can almost hear them saying, "That's not fair!"

Beyond fairness, we can see something that looks like empathy pretty far down the developmental scale. Brosnan went on to describe studies that seem to show that rats have empathy for other rats in their social group. A box is built that can trap one of the rats, and it has a mechanism that other rats outside the box can trigger to free the trapped rat. Researchers found that when a rat sees another rat from its social group trapped in the box, it will work to free the trapped rat without receiving any reward for doing so.

Next the team gave the rats a choice: they were introduced to a testing area that had both a pile of chocolate and one of their trapped comrades. What they found is that the rats would consistently first go and free their friend, and then they would eat the chocolate together. That might be a more highly evolved morality than many of us humans have!

I'm (mostly) joking about that, because I don't actually think rats or monkeys can really be truly morally good or evil in their actions. But in adding the development of fairness and empathy to the major transitions in cooperation I listed in the previous chapter, we can start to see the building blocks of true morality. These developments didn't happen randomly, but in response to stressors in the organisms' environments that had to be overcome, and I think we can say those organisms were participating in the process of moral development, which can be seen over a very long period of time.

I'm not saying single cells were making decisions and thereby participating in their own moral development. But over long stretches of time, the responses to changing environmental conditions clearly became higher levels of cooperation, fairness, and empathy. And the resulting organisms became a new kind of thing that brought us one step closer to beings that had the full capacity for morality.

I've been reading a lot of books on how the proto-morality we find in other species today might have been developing into fuller forms in our ancestors. One of these books is by archaeologist Penny Spikins; it's called *How Compassion Made Us Human: The Evolutionary Origins of Tenderness, Trust, and Morality*. In it, she documents the surprisingly high incidence of ancient human fossils that show evidence of injury or disease. One could look at that and say, "What a difficult and painful world they lived in," and that might be a correct judgment. But it is incomplete.

What Spikins finds when she looks in more depth at these fossils is that many of those people who were injured or diseased beyond their ability to take care of themselves were somehow kept alive for years in those conditions. That doesn't sound like a survival-of-the-fittest strategy, which would have dictated that the weak be discarded. And it might not seem consistent with conferring the kind of reproductive advantage necessary for evolution, until you zoom out further and recognize that the communities that care for the weak are the ones that might be developing other capacities with more direct survival advantages.

I think this understanding of our evolutionary history is really fascinating and ripe for further investigation into how we became the kind of people we are: how our prehuman evolutionary history was necessary for developing moral capacities and how our ancient ancestors responded to the challenges they faced.

For example, the skeleton of a *Homo erectus* female from 1.6 million years ago was discovered in Kenya in 1974. It has an abnormal outer layer of bone, which would have developed over time from heavy bleeding, probably from an excessive intake of vitamin A. For weeks or months she would have suffered from abdominal pain, dizziness, nausea, headaches, blurred vision, and impaired consciousness—hardly the recipe for success on the savanna of East Africa. Yet she did survive long enough for these effects of her disease to become preserved in her skeleton. Someone must have been taking care of her.[3]

This is not some glaring exception to the rule. Fossils have been found from people who had lost their teeth years before their death, thus requiring help in preparing food they could eat. Others have been found with arthritis so severe that it would have prevented independent moving about for years. Such people were not abandoned or left alone but were provided for and almost certainly cared for with compassion.

Why else would Benjamina have been kept alive? She was a *Homo heidelbergensis* child who died between the ages of five and eight, some half a million years ago. Her skull (known as Cranium 14) was found in the Sima de los Huesos site in Spain along with 7,500 fossils from at least twenty-eight different individuals. She had a rare condition called craniosynostosis in which some of the bony plates of the skull fuse together before birth, resulting in a misshapen head and stunted brain growth. Today it is treated with skull-reshaping surgery. Left untreated, pressure builds up as the brain tries to grow, usually leading to severe complications like blindness and seizures. And yet Benjamina received enough care from her community to live five to eight years.

There is evidence that the risky savanna environments our ancestors found themselves in several million years ago put selection pressures on the need to form lasting partnerships, to form parental bonds with their young, and even to fall in love, as "romantic love between mates would ensure that they would stay together and protect their young."[4]

One of the striking aspects of this development of greater emotion and compassion is that it was occurring just as the living conditions were becoming more difficult. The climate was becoming more variable, causing food sources to be less predictable. In this increasingly harsh environment, was it the strongest and meanest who survived? Probably some of them did, taking advantage where they could. But it is out of this pressure cooker that modern humanity emerged.

Spikins shows that difficult conditions in the natural world pushed us to develop emotions and learn to control them, to develop art and appreciate beauty, and to become compassionate. These are the capacities we tend to think of as most uniquely human (at least in their most fully developed forms), and they came about because of challenges. She says,

> When conditions were harshest and survival the most at risk, we see art, beads and ornaments, care for the vulnerable, and the sophisticated and elaborate production of tools. . . . Palaeolithic art is both moving and wonderful, and yet also seems excessive in the context of the societies which produced it. Sensitivity, generosity and care for others seem most obvious precisely when they are hardest to explain.[5]

This lavishness of effort required to create art and care for the vulnerable seems almost irrational to us today when we think of needing to survive in difficult circumstances. Shouldn't we put all our energies into finding the next meal and securing our own little tribe against others who would certainly try to steal from us? That might seem reasonable to us as we project our own cultural assumptions onto the past, but anthropologist David Graeber and archaeologist David Wengrow claim, in their massive book *The Dawn of Everything: A New History of Humanity*, that we are massively wrong to think that way.

They review the statistical frequencies of injuries or disease found in ancient burial sites and what can be determined about how long those people lived after the onset of injury or disease. Based on these, Graeber and Wengrow claim that our conclusions about what life was like should be the exact opposite of Hobbes's state of nature: "In origin, it might be claimed, our species is a nurturing and care-giving species, and there was simply no need for life to be nasty, brutish or short."[6]

Many other books explore the softer and gentler side of our species' evolution. Here are a few examples:

Elliot Sober and David Sloan Wilson, *Unto Us: The Evolution and Psychology of Unselfish Behavior* (Cambridge, MA: Harvard Univ. Press, 1998).

Martin A. Nowak and Sarah Coakley, eds., *Evolution, Games, and God: The Principle of Cooperation* (Cambridge, MA: Harvard Univ. Press, 2013).

Matthew D. Lieberman, *Social: Why Our Brains Are Wired to Connect* (Oxford: Oxford Univ. Press, 2015).

Brian Hare and Vanessa Woods, *Survival of the Friendliest: Understanding Our Origins and Rediscovering Our Common Humanity* (London: One World, 2020).

Anna Machin, *Why We Love: The New Science Behind Our Closest Relationships* (New York: Pegasus, 2022).

The books give a plausible history of the evolution of the kind of behavior I'm calling moral maturity, and they force us to re-think what life was like for our ancient ancestors. Let's not turn the ancient world into Eden; there were still people like Ötzi the Iceman who got shot in the back. But we ought to remember that, as important as it is to reduce violence and increase security, there are other kinds of security worth increasing too. As Graeber and Wengrow note, "There is the security of knowing one has a statistically smaller chance of getting shot with an arrow. And there is the security of knowing that there are people in the world who will care deeply if one is."[7]

The ancient world seemed to be filled with people who cared for each other, and that care came about in the midst of pain and suffering. This recognition moved me closer to responding to the challenge of pain, but I still needed to think more carefully about the theological side of this—how God used pain and suffering (and even death) to move us toward becoming the people God intended us to be. That is the topic of the next chapter.

EVOLVING IMAGE BEARERS

Years ago I heard (or maybe I read it on a T-shirt?) that families who camp together stay together. I don't mean the kind of "camping" where you get an RV that is nicer than your house and "rough it" with air-conditioning and satellite TV. That's not what builds strong families. No, I'm talking about the one-room tent with no electric hookups, no water, and an outhouse for a toilet.

At the time we received this pearl of wisdom, my wife and I had two little boys ages four and one, which is probably not the sweet spot for camping, but we decided to give it a go. We borrowed a tent from some friends and set off for a park in New Hampshire. We found our lovely campsite in the woods by a lake, got the tent set up, and said to each other, "Yes, this is pretty cool. We can see how this would be a good experience for families." And then reality set in.

Our four-year-old just wanted to play in the fire, without having any idea of how dangerous it could be. The one-year-old wandered around the campsite looking for bugs to put in his mouth. Then it rained. Our borrowed tent wasn't as waterproof as we had been promised, and we constantly had to shift things around in the tent so they wouldn't be soaked. The kids cried all night. It was awful.

Driving home, we started talking about how we'd do things differently the next time. Wait . . . what? Next time? Why didn't we vow then and there never to subject ourselves to this again? I don't know. There is something about a challenge, I guess.

We continued camping as a family even as one more kid came along. We faced other challenging circumstances, like poison ivy, food poisoning, and torrential rains. But we continually adapted, attempting to get better. I admit there was one trip when the going got too rough and we gave up and checked into a Courtyard Marriott. Camping wasn't always fun. But looking back now, I see those camping trips as good family experiences. Our family is still close, and often when we're together and reminisce, the conversation turns to those times when we willingly subjected ourselves to less-than-ideal circumstances and lived to tell about it.

Something here gets to the heart of my response to the challenge of pain and suffering. It's not going to make everyone say, "Oh, got it. Now I see why God would create a world that includes pain and suffering." That's setting the bar a little too high, given that our species has been struggling with this problem for as long as we have records. But I do hope to point toward some reasons that might at least make us say, "Hmm . . . that's interesting," and see that an evolutionary response to pain and suffering is at least as good as—and I'll say better than—other explanations for why God would make such a world.

Paul Bloom is a psychology professor at the University of Toronto. He has written a book called *The Sweet Spot: The Pleasures of Suffering and the Search for Meaning*. It is a really interesting discussion of a cluster of topics around suffering, pleasure, satisfaction, and meaning. It addresses the curious fact that some people actually seek out pain—like sexual masochists who seek erotic pain, religious masochists in the Philippines who have themselves crucified during Easter to identify with Christ, or the sporting masochists at the World Sauna Competition who see who can last the longest in 230-degree heat.[1]

Those are extreme examples, however, and more interesting to me is Bloom's overall argument that some degree of suffering is good for us. This is not the easiest thing to perform controlled studies on, but the studies that have been done show that people who have faced little adversity in their lives tend not to deal well with pain. Those who have faced a moderate amount of adversity are the best equipped to handle pain themselves and to be compassionate toward others who are suffering.[2]

Author Rebecca Solnit has investigated tragedies more anecdotally in her book *A Paradise Built in Hell: The Extraordinary Communities That Arise in Disaster*. "It was the joy on their faces that surprised me." That's how Solnit described talking to lots of different people about the tragedies their communities had endured—from 9/11 to Hurricane Katrina, from Canadian ice storms to oppressive heat in India. The disasters shouldn't be desired, she dutifully notes, but neither should their side effects be ignored.

> *The desires and possibilities awakened are so powerful they shine even from wreckage, carnage, and ashes. What happens here is relevant elsewhere. And the point is not to welcome disasters. They do not create these gifts, but they are one avenue through which the gifts arrive. Disasters provide an extraordinary wisdom into social desire and possibilities, and what manifests there matters elsewhere, in ordinary times and in other extraordinary times.[3]*

For the last several years, I've followed the World Happiness Report, an annual survey of countries around the world. I have looked longingly at Finland, Iceland, Norway, and those other Northern European countries that always end up at the top of the list. They have more socioeconomic equality, lots of social services, and the highest GDP per capita. Their citizens do not suffer much. But interestingly, Bloom reveals that neither do they score very high on a different list: the most *meaningful* lives. The countries at the top of that list are Sierra Leone, Senegal, Laos, and Cuba—places with little wealth or security and a comparatively low GDP. These are places where life is a struggle.[4]

Why is it that life in difficult countries, or even tragedies in otherwise easier places to live, gives rise to joy and meaningfulness that go beyond the happiness born of security? I find it hard to believe that other animal species can find meaningful lives in more challenging circumstances that don't lend themselves to natural flourishing. I think we humans can do it because we have evolved the capacity to think about things differently, to "see as."

We live in the very different world of the soul, as I discussed in part IV. Our mood and even our sense of well-being are not just a function of the things that happen to us, but also—and perhaps more importantly—how we understand and respond to the things that happen to us. This is the central insight that helped me make sense of the history of pain and suffering.

The New Testament book of James begins with this very insight: "My brothers and sisters, whenever you face various trials, consider it all joy, because you know that the testing of your faith produces endurance" (James 1:2–3). We are able to see trials *as* an occasion for joy when we believe they are good for us in some way. And for the response I'm making to the challenge of pain, it is interesting that James continues in the next verse, "Let perseverance finish its work, so that you may be mature and complete, not lacking anything" (James 1:4, NIV).

Maybe James wasn't talking about moral maturity, and I definitely don't think he had our evolutionary development in mind. But neither do we find any hint of the sentiment that a good God would spare us from trials. Rather, the trials are understood as doing something to us: producing endurance, which leads to maturity.

Nobody is suggesting that suffering is good in and of itself. And I don't think most people, even when given the statistics on happiness and meaning, would choose to live in places where the likelihood of suffering is higher. But at the same time I'm finding it increasingly persuasive that suffering, adversity, and struggle can do something to us that might be beneficial. Could it be that God intentionally placed our ancestors—even the prehuman ones—in an environment that would induce some pain and suffering so they would develop the capacities of cooperation, empathy, and love?

We're pressing up against the limits of what can be said with much confidence. But it's worth speculating a little bit within the parameters of what we know about the natural history of our species through science and what we know about our calling and vocation through theology. According to science, beyond all reasonable doubt, *Homo sapiens* evolved from earlier hominins around 300,000 years ago and have common ancestry with all other life on Earth. From traditional Christian and Jewish theology, human beings were created in the image of God and so given a special role to play in creation as God's representatives. The question is how we combine these two facts into one coherent story.

Back in chapter 13, I discussed the possible relationships between *Homo sapiens* and humans. My slightly different question now is this: when in our evolutionary development did we become God's image bearers to the rest of creation? Here too there are options.

For some people, the gradualness of evolution and the relatively smooth transitions described in part III are problematic. They want to see a definitive break between those who do not bear God's image and those who do. Can we speculate about how that might have happened without transgressing the parameters laid down by science and theology?

Let's say that evolution was occurring as described by scientists until there was a population of hominins that developed the capacity for moral maturity, which I take as necessary to be God's representatives. I'm not equating the capacity for moral maturity with the image of God; rather, I'm saying that to fulfill that calling our species needed to have moral maturity. In this view, becoming image bearers depended on God's call, and that happened all at once in a moment in time when God entered into a new relationship with our species.

Maybe that was 300,000 years ago as our species became anatomically modern; maybe it was 60,000 years ago as our species was becoming behaviorally modern. It could have even been 10,000 years ago to fit closer with the Adam and Eve story and its aftermath in Genesis. Or it could have been 600,000 years ago before our ancestors split with the ancestors of Neanderthals (assuming the necessary capacities had developed by then). The point here is that the capacities needed to properly fulfill the vocation of image bearers had to have been in place as a first step, and then God was revealed in a new way to them. They bore the responsibility from that point onward of imaging God and stewarding creation.

I think that's a defensible position to hold, and it falls easily within the parameters set down by science and theology. But I don't think it is the only such position. I'm not persuaded that conferring the image of God and our taking on that responsibility had to be an instantaneous or punctiliar event.

In this part, I've relied on there being an analogy between our species and us as individuals. I think it holds for this point too. Think of a baby growing up to become an adult. This is a gradual event for which it is difficult to draw nonarbitrary lines of development. But we clearly hold twenty-one-year-olds responsible for things that we do not hold sixteen-year-olds responsible for. And the same goes for ten-year-olds and three-year-olds. Of course there may also be individual differences determining when someone is ready to handle a responsibility like staying home alone for a few hours or overnight. I also wouldn't say that the actions of an infant carry moral responsibility yet, but it isn't clear exactly when they start to do so. It doesn't seem right to say that a child is morally responsible one day when they weren't the previous day.

In the same way, we might say that our species grew up into the responsibility of stewarding creation—not all at once, but over time. Perhaps God had been revealing this vocation to us all along, and as our capacities developed, we came to understand more and more of that calling. And God held us increasingly responsible at developmentally appropriate intervals. In this account, we wouldn't have been image bearers one day but not the day before. I don't see anything in this option that falls outside the parameters set down by science and theology.

So far so good, but one more aspect of my response to the challenge of pain and suffering still needs to be addressed. And I'm going to need another guide to work through it.

FINDING DEEPER FAITH
WITH SIMONE WEIL

We can begin to understand that the world is the way it is because it was intended to make us the way we are. Our ancient ancestors' responses to difficult circumstances helped form the capacities that would eventually become moral maturity in us. But I haven't yet properly grappled with the implication of this, namely that God created a world that had those difficult circumstances already in it. Must we say, then, that God created pain and suffering?

Long before humans were around and committing the kind of morally reprehensible acts that could be called sin, the world operated on evolutionary principles according to which some individual organisms suffer and die painful deaths. Hurricanes caused pain and suffering, viruses caused disease, cancer destroyed from the inside out, and predators with sharp fangs caught their prey and ate them alive.

At best, we have to ask how this kind of creation could be called good; at worst, we have to wonder whether it makes God the author of evil. I'll say it again: don't expect a conclusive argument here that solves the problem of pain and suffering once and for all. But maybe some of these considerations can help.

For one European trip, I arranged my schedule so I had an overnight stay in London and an afternoon departure the next day. That gave me time to take a short train ride south to the city of Ashford. From the station I walked about a mile and a half to the Bybrook Cemetery where Simone Weil's gravesite is located. She was buried there in 1943, having died at the tragically young age of thirty-four while working for the French Resistance during World War II.

Weil has become known as a Christian mystic and spiritual master, but ethnically she was a nonpracticing Jew. She studied philosophy and became enamored with Christianity through a friendship with a priest, without ever officially converting. She wrote prolifically but published no books during her lifetime.

I'm not sure why I wanted to see her grave. I suppose I had something of the same impulse as the guy in chapter 17 I saw genuflecting before the little bone chip of Saint Gregory. There were a lot more of her bones there, six feet below the ground, but nothing magical happened when I went near them. I paid my respects, took a couple of selfies, and then took the train back to London.

Weil's intellectual remains are in several books that were compiled by her friends from the notes and letters she left behind. Her thinking is complex and not always consistent. I find it best to read her work as flashes of insight that were never quite woven into coherent arguments. This of course is a reflection of scattered notes rather than a finished book. But I think it might also be a function of the fact that she, more than most philosophers, lived her philosophy in the messiness of the real world, rather than stitching together abstract ideas in an ivory tower. Her ideas about the plight of the working class were refined by the mind-numbing work she did in factories. She didn't simply theorize about the oppressed people in Spain's civil war but went to join their resistance movement. And her thinking about pain and suffering occurred alongside the debilitating migraine headaches that she had all her life.

Central to her understanding of pain and suffering is Weil's claim that the act of creation—of giving existence to something else—must be understood as God withdrawing the divine being from that creation. That might sound strange or even heretical, but she doesn't mean it in the sense that God abandons creation. It is more like the idea of *kenosis* found in the New Testament book of Philippians: God "emptied himself" (that's how most English versions translate the *kenosis* passage) to become human in the person of Jesus Christ (Phil. 2:7). But according to Christian theology, Christ was still God—just now with certain limitations that go with being human. How much more emptying must it take to give existence to something that is not God?

God is supreme goodness, and created things are not God (to

say otherwise would be heretical!). So according to Weil, giving created things their own existence has to make them something other than the absolute goodness of God. It shouldn't surprise us, then, to find suffering and pain in the created order. But that doesn't mean God has no love for creation—just the opposite, in fact. In an essay called "Some Thoughts on the Love of God," Weil wrote:

> *It was by an inconceivable love that God created beings so distant from himself. . . . The evil which we see everywhere in the world in the form of affliction and crime is a sign of the distance between us and God. But this distance is love and therefore it should be loved. This does not mean loving evil, but loving God through the evil.*[1]

I'm still not sure we should call the difficult and painful parts of creation "evil" when they result from natural processes. The lightning that burned down the house (recall the thought experiment in chapter 22) was tragic, but I don't think it's morally evil in the same way that the arsonist is evil. Weil's point, though, is that these difficult parts of creation should be seen as signs of God's love. To love is to will the good of another thing, and for God to create another thing to love, it would have to be distinct from God and therefore less than the absolute goodness of God.

This is some sophisticated and tricky metaphysics from Simone Weil, and it raises all kinds of questions to which we won't find definitive answers in her writing. She's trying to sort out how a good and loving God can create a world with pain and suffering. Her account suggests that suffering and love are intimately entangled, just as they were in the ultimate example she cites of suffering and love: Christ on the cross.

The cross of Christ also points to the fact that creation was not intended to stay in this state forever. Remember that God didn't create things initially the way they were ultimately intended to be. In this sense too we might understand how the initial creation could be pronounced "good" even though it wasn't yet the good it was supposed to become. Think of this like a baby who is pronounced by the doctor at her six-month checkup to be healthy and in very good shape. But she doesn't have any teeth yet, she can't walk, and she can't talk; if she is still in that condition at her six-year checkup, the doctor will probably give a very different assessment of her condition.

When we look at the world today and see all the pain and suffering, our assessment of it as "not good" is because we know that it should be something different by now. Its initial condition was good for that stage of its development—it produced an astounding lavishness of life as well as the capacities in us for moral maturity. But it wasn't intended to stay that way.

According to Christian theology, the cross of Christ and his resurrection inaugurated a new era, and it is from this vantage point that we can recognize all creation as groaning, the way it is described by Paul in Romans 8. We should look at that passage to see how well it fits with this Weil-inspired description of creation.

The Apostle Paul was engaged in a project of reconciling his experience of the risen Christ with the sacred texts of his religious tradition—what Christians today call the Old Testament. This often included rereading those texts in the light of what he came to know. In rereading texts, you can't make them mean something completely different, but you can find new resonances that speak to current circumstances that perhaps weren't explicitly intended in the original. That is what I'm attempting to do—not with the authority of an apostle, but as one who is speculating about how we might see the natural history of evolution in the light of Christian faith.

In this regard, I'm particularly interested in Paul's description of suffering and creation in Romans 8:18–23:

> *I consider that the sufferings of this present time are not worth comparing with the glory about to be revealed to us. For the creation waits with eager longing for the revealing of the children of God, for the creation was subjected to futility, not of its own will, but by the will of the one who subjected it, in hope that the creation itself will be set free from its enslavement to decay and will obtain the freedom of the glory of the children of God. We know that the whole creation has been groaning together as it suffers together the pains of labor, and not only the creation, but we ourselves, who have the first fruits of the Spirit, groan inwardly while we wait for adoption, the redemption of our bodies.*

To use the same metaphor Paul uses in this passage, I find this description pregnant with meaning. Creation was "subjected to futility." I'm not saying Paul knew about the millions of years of suffering and death on Earth before humans were around to sin. I simply mean that there are ways of describing—ways of seeing the natural history of our planet—that resonate with the grand themes of this Gospel message.

God created a world where there are predators, hurricanes, viruses, and the like that are capable of causing suffering and death. God said the world was good, even though it also needed to be filled and subdued. God created it that way on purpose. Why? To give us something to do? Maybe . . . but maybe also because it would do something *to* us. It would train us to become the kind of people God ultimately intended. Maybe God created a world that needed subduing, a world that was subject to futility, for the specific reason of creating a particular kind of life—life that could become morally mature.

Furthermore, the world was created this way so that it too would become something else. Here is the pregnancy metaphor: "We know that the whole creation has been groaning in labor pains until now." This suggests that creation is in a gestational period, that it has not yet been born into what it is ultimately intended to be. Remember from part II that God didn't create the world originally as it was ultimately intended to be.

When will it become that? Back up three verses: "For the creation waits with eager longing for the revealing of the children of God" (8:19). When we children of God finally live up to the calling we were given as image bearers and stewards, then creation too will become what it was ultimately intended to be. Because we didn't do that originally, and introduced sin into the world instead, we needed the atoning work of Christ to make things right. And although that has already happened, we await the further action of God for "adoption, the redemption of our bodies." That is Paul's allusion to the final resurrection that will consummate the Kingdom of God in the new heavens and new Earth. That's when there will be no more "natural evils."

Creation was subjected to futility not as a punishment but "in the hope that it would be set free." God seems to have a purpose in mind for subjecting creation to futility, and that purpose seems connected to our role as the children of God, as God's image bearers. In a similar vein (though on a much smaller scale), think of a parent who gives a child a chore to do that the child may find irritating or unpleasant: the parent's long-term goal is to help the child grow to be responsible, thoughtful, and able to see a task through to completion. So too God had our ultimate good in mind when the world was created as it was.

God created a place that needed to be subdued, a place where the image bearers would have a job to do. We are supposed to have an influence on creation, to cooperate with God in setting it free. But then, subjecting creation to futility (which includes, on this reading, allowing things like hurricanes and viruses) may have had a purpose for our own development too. God couldn't create morally mature beings from scratch because moral maturity demands participation in the process. The scientific details we've uncovered about the natural history of *Homo sapiens* seem to resonate with this story.

This account won't be persuasive to everyone. It is not the solution to the problem of evil. None of us has that. It is too wondrous for us to know, in the words of Job; it is above our pay grade. But Simone Weil has persuaded me that something can be said that at least points in the direction of reconciling pain and suffering, which science tells us have existed from the beginning, with God's love and intentions for us and all of creation.

Weil wrote in one of her notebooks, "The extreme greatness of Christianity lies in the fact that it does not seek a supernatural remedy for suffering but a supernatural use for it."[2] If God simply

intervened to eliminate suffering, that would be the same as elim-
inating our existence. So then, "those who ask why God permits
affliction might as well ask why God created."[3]

According to this perspective, I think it is legitimate to say that all
of creation has a cruciform nature: love and suffering penetrate to
the core of created things, pointing to the cross of Christ. Through
the twin lenses of evolutionary science and Christian theology, we
can see that, by creating the world, God allowed pain and suffering
to exist, and then God used pain and suffering to shape us into image bearers so that we might participate in transforming the groaning creation into what it was ultimately intended to be.

CONCLUSION

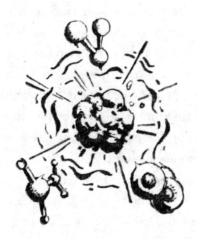

At a recent conference for spiritual directors, I spoke on a panel about the *Language of God* podcast. Also on that panel was Cortland Dahl, who is a research scientist and chief contemplative officer for Healthy Minds Innovations, which has developed a popular mindfulness app called Healthy Minds. From Cort's title, you might not be surprised to hear that he practices Buddhism. His is not a casual or faddish Buddhism, either. He spent more than a decade studying in Nepal.

We hit it off pretty well in our short time together at the conference, so I downloaded his app and have used it occasionally since then. It doesn't adopt any particular religious perspective but instead aims to promote mindfulness in a way that is applicable to the mental health of us all. I've been curious how I might apply the principles and practice of mindfulness within my own religious faith.

While finishing the manuscript for this book, I was driving my once-a-week one-hundred-mile commute, and I listened to a meditation from the app about compassion and developing empathy for others. I thought that was interesting (and perhaps providential?) as I was deep in the process of trying to sort out the chapters on suffering and the development of moral maturity in our species. One of the exercises was to envision people you interact with and to think about them, "May you be free from suffering and hardship." For me, that naturally turns into a prayer, but then after thinking some more about how I've presented the key role that suffering has played in our development—both as a species and as individuals—I wondered if that was really what I wanted to pray for the people in my life. Of course I don't want their lives to be hard, but I do want them to be meaningful. Should I pray instead that they have just the right amount of suffering?

I ended up emailing Cort to ask about this. Not only is he a Buddhist, but he also has a PhD in psychology, and I figured he'd know about Paul Bloom's work and the book I discussed in chapter 25. I asked him, "I'm curious whether you've engaged with this research at all, and whether it is more important to feel happy or to feel a sense of meaning and purpose?"

He answered me with a lengthy email that concluded, "Summing up a great deal of research, I think we could safely say that being concerned about, and willing to help alleviate, the suffering of others is one of the most meaningful things we can do. Science and the world's wisdom traditions seem to largely agree on that point."

Of course that is right. We should not seek out suffering (perhaps with the exception of camping!). Even Jesus prayed that the "cup" of suffering would pass from him. And we shouldn't pray that our loved ones find meaningful lives by suffering. Instead, let's pray that they find meaningful lives by alleviating the suffering of others.

There is a curious tension here: some degree of suffering appears to be good for us, but we don't want it and shouldn't seek it out. The case I tried to make in the previous chapters was that suffering has been an important driver for our becoming the kind of creatures we are now and that this is more consistent with the Christian theological tradition than I had been led to believe by my conservative community. That doesn't mean, however, that we should simply accept suffering as inevitable. It's not inconsistent to think that suffering could be used for good in the past, but that we should work to relieve suffering in all of creation in the present. And those of us who are Christians believe that a future is coming in which there will be no more suffering.

Really? No more suffering? What does science have to say about that? Throughout this book I've attempted to let science and theology be in dialogue with each other, and I've claimed that this has led to a deeper, more relevant faith. Have we reached an impasse here, though? A place where science and Christian theology must diverge, forcing us once again to choose one or the other? I hope not. This sounds like it could be another challenge that takes several more chapters to unpack and respond to. But talk about the future is pretty speculative, so I'll just gesture toward some possible solutions in this conclusion.

Let's start by recapping the story as I've told it thus far with some of the central insights that emerged from the various challenges encountered along the way. God intentionally created us to be image bearers, but not even God can create morally mature people without their cooperation in the process (the challenge of pain). This means that God didn't create things originally the way they were ultimately intended to be (the challenge of species). Instead, God took a long time, not being concerned about time and efficiency (the challenge of time), and let evolution produce the right kinds of body types that could give rise to the right kinds of souls, ultimately allowing language to develop, along with the ability to see things *as* other things (the challenge of soul). In spite of this important intention for humanity, creation certainly isn't all about us; God created in a way that would bring us about but also so that an extraordinarily lavish variety of other life could exist; God delights in the small things too and loves to see them flourish (the challenge of time). Ultimately, according to the sacred texts God took up into service in my religious tradition (the challenge of the Bible), we did become God's image bearers, and now we have a special duty to care for this groaning creation until God acts again to bring about the eternal Kingdom of Heaven (the challenge of pain).

It's that last part that makes some of my scientifically oriented friends nervous. Looking back at history, they admit, we might be able to reconceptualize what science has figured out, seeing it *as* the hand of God at work. But talk of some future eternal state seems to go against scientific principles and ends up appealing only to the miraculous. The sun is going to burn out in a few billion years. Perhaps by then we'll have figured out how to transport ourselves to other star systems, but that too is only delaying the inevitable. Stars will stop forming in our galaxy after 40 or 50 billion years, and after 100 million years the other galaxies will be so far away that their light will never reach us. And worst of all, in 10^{31} years, all the protons and neutrons in the universe will have decayed, leaving behind no trace of anything that has ever existed.[1]

That last number—10^{31} years—represents a very long time, but it's just the blink of an eye from the perspective of eternity. Is there a way of reconciling faith's view of eternity with what we now know scientifically? No, not if we mean to show that both of these visions of the future are true. Our current science shows pretty definitively that our universe is destined for a temporary existence. But maybe we can take some solace in the fact that we don't know everything now. There are still so many unknowns and mysteries from the perspective of science, and I'd suggest caution in prophesying about the end-times in much detail from the perspective of faith. However, I think we can at least gesture toward an eternal future in a way that doesn't contradict what we know from science now. Here is a glimmer of hope that the doom and gloom of today's science may not be the last word on our ultimate future.

There have been three transformative moments in the history of our universe, sometimes called the three big bangs.[2] The first is what is normally called the big bang: the transition from nothing at all to the existence of matter and energy (maybe there was something like a multiverse from which our universe sprang, but that only pushes back the problem to the origin of multiverses). The next big bang occurred when some of that matter and energy became alive. And then the third took place when some life became conscious or self-aware.

The existence of consciousness (what I've called mind or soul) somehow depends on life, and life somehow depends on matter and energy. We don't yet have good scientific explanations for how these transitions occurred, and maybe we never will. But I'm not proposing they can only be explained by a miraculous intervention by God. I'm simply saying that very different kinds of existence emerge from the lower levels in ways that are utterly surprising when considering only what came before them. Scientists who only knew about matter and energy would be surprised by the emergence of life. (Yes, I know those scientists would have to be alive to be surprised. It's just a thought experiment, so play along!) And similarly, scientists who only knew about single-celled and other simple life would be amazed that consciousness could eventually emerge from that. So what's to stop us from speculating about some further kind of existence that could emerge from consciousness and soul? Why should we think that there could only ever be three such transitions? Maybe in the future some of the conscious life will transform or emerge into a different kind of being that transcends the limitations of the world as we know them now. The Apostle Paul said that after resurrection we will have "spiritual bodies" (1 Cor. 15:44). Maybe the fourth big bang is the emergence of spirit. Perhaps that kind of existence can live for eternity in the loving Kingdom of God.

What does that do to our dialogue between science and faith? Does that mean faith triumphs over science? I don't think so. The scenario I suggest wouldn't have to signal the end of science. Rather, we could believe that the emergence of spirit would be just as amenable to future scientific study as matter, life, and consciousness are today. Perhaps in the eschaton, science will flourish as never before, as spirit-enabled scientists can study more ultimate laws of nature than those we have access to in our present state.

Can I prove that any of that is true? Nope. But neither do I see how any of it can be proven untrue through the methods of science. Faith is too often taken to mean believing without evidence. A better definition of faith is commitment without certainty. In that sense, it is no less an act of faith to commit oneself to naturalistic explanations that can't ultimately be proven to tell the whole story, than it is to commit oneself to there being something supernatural at work in the universe.

I'm aware that not everyone will see things the same way. My primary objective here has not been to convince everyone that they should believe just like I do. I'm simply trying to show that the way I have reconciled evolution and Christian faith is reasonable in the light of evolutionary science. And even further, I think I've shown that the kind of faith I've landed in is a deeper, more authentic version of faith for our world today. But I'm aware that I cannot demonstrate this with certainty. I'll simply note that I hope it's true, and I'll conclude by saying a little more about hope.

On a *Language of God* episode, Stanford neuroscientist Bill Newsome gave a fascinating characterization of his own religious faith: "I often say that I think religious belief is about one-third cognitive assent and about one-third intuition and about one-third sheer, unadulterated hope."[3]

Our faith does involve a cognitive component. We believe things and produce evidence for them. Intuition is what seems reasonable to us, given the rest of our beliefs and experiences, and of course this can vary widely from person to person. Then the rest is sheer, unadulterated hope.

The older I get, the less I think I know—at least with certainty—and the more I hope. I hope there is a God who loves us and all creation, desiring and working for our ultimate and eternal good. That is what I want to be true, but I don't think hope in this sense is just wishful thinking. Legitimate hope is grounded in a coherent and reasonable story of how things might go.

That's what I've tried to do in this book. I've followed the sacred chain from the first moments of creation, through the lavishness of life that developed, and on to the evolution of our own species. I've not sacrificed the scientific understanding of these events to preserve my religious faith. Quite the opposite, in fact. Allowing evolution to be a dialogue partner has led me to a deeper faith. I've found that science and faith together can tell a coherent and reasonable story about our past, present, and future. That gives me hope that the story could really be true.

ACKNOWLEDGMENTS

The journey I describe in this book was not made alone. I mentioned a number of guides I benefited from along the way in my "Finding Deeper Faith" chapters, but I never actually met any of them in the flesh (except the redwoods). I've had to imagine them accompanying me at various segments of the journey. But there were others who were actually there, supporting and encouraging me in important ways. I'll start the roll call where the book itself began.

The journey I've described started in earnest with long conversations with my fellow philosophy professor Chad Meister at the college where we taught. We endeavored to understand our faith in the light of evolution—and to understand evolution in the light of our faith. We drank a lot of coffee and talked behind closed doors for many years about these issues and how our community might receive them. He too is now a former professor.

When I had to leave the college, I was grateful to be able to work and earn a living at BioLogos. But full-time jobs at small nonprofit organizations usually come with a catch: I'd need to find a grant to help support my work. That ended up coming from the John Templeton Foundation in the form of a multidisciplinary exploration of what it means to be human. I'm very grateful to Jeff Schloss for steering me toward this topic and suggesting a lot of books that became foundational for my research. That led to a big literature review, and for that I'm thankful for the work of my intern, Ryan Bollier. I still look back at the précis he wrote when I'm trying to remember an important point from one of those books. And thanks to Fritz Hartman, the library director at the Harold and Wilma Good Library at Goshen College, for giving me access to their books and even tracking down some they didn't have.

All of the previous books I've written or edited were published by academic presses. It is a journey in and of itself to migrate from those to a publisher like HarperOne, and I must acknowledge the help I had along the way. Beth Graybill encouraged me over coffee more than once, making me believe that I could and should do this. Nancy Erickson made some important connections for me, and I'll be buying her coffee for quite some time. Francis Collins hooked me up with his literary agent, Gail Ross, who took a chance on me. Then she and her staff at the Ross Yoon Literary Agency honed and shaped the sprawling ideas I brought until they were suitable for the general public. Gail opened doors for me at publishers, and I was thrilled this project landed with HarperOne. My editors, Katy Hamilton and Gabi Page-Fort, deftly guided this project through all the publishing hoops, and their team improved it at every step in the process.

Several of my colleagues at BioLogos also need to be acknowledged for their support and encouragement. Deb Haarsma and Laura J. Landmann made this project possible for my work life. Most significantly in that regard was supporting my sabbatical request and covering for me while I traipsed around Europe for six weeks and then huddled over my computer writing the manuscript for ten more.

My colleague and friend Colin Hoogerwerf read the entire manuscript and suggested some helpful changes. More importantly, though, were the ideas that came out of work we did putting podcast episodes together and from the stimulating conversations we've had on many work trips—may there be many more!

My friend and former colleague Kathryn Applegate also read the entire manuscript and suggested a lot of edits. Any sentences that are still unclear or wonky are almost certainly because I didn't accept her changes.

Finally, I want to acknowledge my family, which brings a lump to my throat because of how important and valuable they have been to me. This starts with my parents, Ron and Nancy Stump. They too were born into the conservative Christian community my family was part of, and if everyone in that community were like them, I'd probably still be there. They have been models of integrity all my life and didn't shut down questions. My decision to go to graduate school for philosophy was a curious and maybe even worrying choice to them, but they have been unflaggingly supportive throughout all the twists and turns my life has taken.

My kids and their partners—Casey, Trevor, Sloan, and the two Emilys—have had a bigger influence on my ideas than they could know. Sloan is responsible for all of the fantastic illustrations in this book (which I can say without hubris because she gets none of her artistic ability from me); check out her website at sloanstump.com.

My good wife Chris is closest and dearest to me of all the people on the planet. Our life together has been a grand adventure, and she has been an amazing traveling companion (besides being my biggest fan and a constant source of strength and encouragement).

Finally, I'm so proud of the next generation of Stumps: Finley and Auden. They are so much fun and such an inspiration. I dedicated the book to them in the hope that they have long lives ahead of them in which their encounters with science—be they many or few—are wondrous and faith-deepening.

NOTES

CHAPTER 1: COMMUNITIES OF ORIGIN

1. Jon D. Miller, Eugenie C. Scott, and Shinji Okamoto, "Public Acceptance of Evolution," *Science* 313, no. 5788 (August 2006): 765–66, https://www.science.org/doi/10.1126/science.1126746.

2. Jon D. Miller et al., "Public Acceptance of Evolution in the United States, 1985–2020," *Public Understanding of Science* 31, no. 2 (2021): 223–38, https://doi.org/10.1177/09636625211035919.

3. Lee Rainie et al., "Chapter 4: Evolution and Perceptions of Scientific Consensus," Pew Research Center, July 1, 2015, https://www.pewresearch.org/science/2015/07/01/chapter-4-evolution-and-perceptions-of-scientific-consensus/.

4. Cary Funk et al., "Biotechnology Research Viewed with Caution Globally, but Most Support Gene Editing for Babies to Treat Disease," Pew Research Center, December 10, 2020, https://www.pewresearch.org/science/2020/12/10/biotechnology-research-viewed-with-caution-globally-but-most-support-gene-editing-for-babies-to-treat-disease/.

5. Edward Larson, *Summer for the Gods: The Scopes Trial and America's Continuing Debate over Science and Religion* (New York: Basic Books, 1997), 6.

6. "Walking with Stan Rosenberg," *Language of God* podcast, October 13, 2022, https://biologos.org/podcast-episodes/walking-with-stan-rosenberg-parks-road-oxford.

CHAPTER 2: CREATIONISM

1. John C. Whitcomb and Henry M. Morris, *The Genesis Flood: The Biblical Record and Its Scientific Implications* (Grand Rapids, MI: Baker Book House, 1961), 118.
2. Genesis Apologetics, "Debunk Evolution Classroom Promo Video," YouTube, https://www.youtube.com/watch?v=3IbhemNg1yg.
3. To search for common ancestors of species, visit TimeTree.org.
4. "An Elaboration of AAAS Scientists' Views," Pew Research Center, July 23, 2015, https://www.pewresearch.org/science/2015/07/23/an-elaboration-of-aaas-scientists-views/.

CHAPTER 3: FINDING ANSWERS

1. "About," Answers in Genesis, accessed October 26, 2022, https://answersingenesis.org/about/.
2. "Resolved!," Creation Museum, January 3, 2013, https://creationmuseum.org/blog/2013/01/03/resolved/.
3. Hannah Ritchie, "How Many Species Are There?," Our World in Data, November 30, 2022, https://ourworldindata.org/how-many-species-are-there.
4. Joel Duff, "The Young-Earth Hyper-evolution Hypothesis: A Collection of Critiques," Naturalis Historia, August 3, 2016, https://thenaturalhistorian.com/2016/08/03/the-young-earth-hyper-evolution-hypothesis-a-collection-of-critiques/.
5. Kathleen C. Oberlin, *Creating the Creation Museum: How Fundamentalist Beliefs Come to Life* (New York: New York Univ. Press, 2020).
6. Brad Kramer, "My Trip to the Creation Museum," BioLogos, August 17, 2015, https://biologos.org/articles/my-trip-to-the-creation-museum.

CHAPTER 4: MY CHURCH LIED TO ME

1. "Philip Yancey: What Good Is Disappointment?," *Language of God* podcast, April 11, 2019, https://biologos.org/podcast-episodes/philip-yancey-what-good-is-disappointment.
2. Stephanie Kramer et al., "How US Religion Composition Has Changed in Recent Decades," Pew Research Center, September 13, 2022, https://www.pewresearch.org/religion/2022/09/13/how-u-s-religious-composition-has-changed-in-recent-decades/.
3. Jeffrey M. Jones, "Belief in God in U.S. Dips to 81%, a New Low," Gallup, June 17, 2022, https://news.gallup.com/poll/393737/belief-god-dips-new-low.aspx.

4. The Fuller Youth Institute has produced lots of resources about this. A good place to start is "I Doubt It: Making Space for Hard Questions," by Kara Powell and Brad M. Griffin, March 10, 2014 (https://fuller youthinstitute.org/blog/i-doubt-it) and *Can I Ask That? Eight Hard Questions About God and Faith* by Jim Candy, Brad M. Griffin, and Kara Powell (Pasadena, CA: Fuller Youth Institute, 2014).
5. Ted Davis, "Science and the Bible," BioLogos, November 1, 2019, https://biologos.org/series/science-and-the-bible.

CHAPTER 5: FINDING DEEPER FAITH WITH C. S. LEWIS

1. C. S. Lewis, *Reflections on the Psalms* (San Francisco: HarperOne, 1958), 130.
2. Lewis, *Reflections on the Psalms*, 129.

CHAPTER 8: GOD'S PRIORITIES

1. Margaret Osborne, "An Estimated 20 Quadrillion Ants Live on Earth," *Smithsonian*, September 21, 2022, https://www.smithsonianmag .com/smart-news/an-estimated-20-quadrillion-ants-live-on-earth -180980804/.
2. Jason Daley, "Humans Make Up Just 1/10,000 of Earth's Biomass," *Smithsonian*, May 25, 2018, https://www.smithsonianmag.com/smart -news/humans-make-110000th-earths-biomass-180969141/.
3. "This is Not a Typo: One in Four Animals Known to Science Is a Beetle," *Short Wave* podcast, NPR, September 10, 2020, https://www.npr .org/transcripts/910447004.
4. Ashley Hamer, "99 Percent of the Earth's Species Are Extinct—But That's Not the Worst of It," Discovery, August 1, 2019, https://www .discovery.com/nature/99-Percent-Of-The-Earths-Species-Are-Extinct.
5. pkrumins, "Carl Sagan: If You Wish to Make an Apple Pie from Scratch, You Must First Invent the Universe," YouTube, https://www.youtube .com/watch?v=7s664NsLeFM.

CHAPTER 9: RETHINKING TIME AND THE BIBLE

1. "Richard Middleton: Interpreting Biblical Genealogies," *Language of God* podcast, September 16, 2021, https://biologos.org/podcast -episodes/richard-middleton-interpreting-biblical-genealogies.
2. "John Walton: More Than History," *Language of God* podcast, September 19, 2019, https://biologos.org/podcast-episodes/john-walton -more-than-history.

CHAPTER 10: FINDING DEEPER FAITH WITH ARUNDHATI ROY AND G. K. CHESTERTON

1. Arundhati Roy, *The God of Small Things* (New Delhi: Penguin Books India, 2002), 32–33.
2. "Makoto Fujimura: Creating Beauty from Brokenness," *Language of God* podcast, March 31, 2022, https://biologos.org/podcast-episodes /makoto-fujimura-creating-beauty-from-brokenness.
3. G. K. Chesterton, *Orthodoxy* (New York: John Lane, 1909), 108–9.

CHAPTER 11: WHAT IS A HUMAN?

1. Quoted in Londa L. Schiebinger, *Nature's Body: Gender in the Making of Modern Science* (New Brunswick, NJ: Rutgers Univ. Press, 2004), 80.

CHAPTER 12: OUR CLOSEST COUSINS

1. The skull of a child was discovered in Belgium in 1829. It is known as Engis 2 and has since been judged to be Neanderthal, making it the first known discovery. But it did not figure as prominently in the discussion of what these creatures were.
2. James Walker, David Clinnick, and Mark White, "We Are Not Alone: William King and the Naming of the Neanderthals," *American Anthropologist* 123, no. 4 (December 2021): 806, https://anthrosource .onlinelibrary.wiley.com/doi/pdf/10.1111/aman.13654.
3. Monty White, "The Caring Neandertal," Answers in Genesis, September 1, 1996, https://answersingenesis.org/human-evolution/neanderthal /the-caring-neandertal/.
4. Clive Finlayson et al., "Late Survival of Neanderthals at the Southernmost Extreme of Europe," *Nature* 443 (September 2006): 850–53, https://doi.org/10.1038/nature05195.
5. Walker, Clinnick, and White, "We Are Not Alone," 808.
6. Bridget Alex, "Neanderthal Brains: Bigger, Not Necessarily Better," *Discover*, September 21, 2018, https://www.discovermagazine.com /planet-earth/neanderthal-brains-bigger-not-necessarily-better.
7. Ian Tattersall, *Masters of the Planet: The Search for Our Human Origins* (New York: St. Martin's Griffin, 2012), 168.
8. Emily Willingham, "Genes Linked to Self-Awareness in Modern Humans Were Less Common in Neandertals," *Scientific American*, May 10, 2021, https://www.scientificamerican.com/article/genes -linked-to-self-awareness-in-modern-humans-were-less-common-in -neandertals.

9. Ian Tattersall, *Understanding Human Evolution* (Cambridge: Cambridge Univ. Press, 2022), 123.

10. Tattersall, *Understanding Human Evolution*, 126.

CHAPTER 13: OUR ANCESTRY IN BASEBALL CARDS

1. Richard Dawkins, *The Magic of Reality: How We Know What's Really True* (New York: Free Press, 2011), 38–40.

2. Toshiko Kaneda and Carl Haub, "How Many People Have Ever Lived on Earth?," Population Reference Bureau, November 15, 2022, https://www.prb.org/articles/how-many-people-have-ever-lived-on-earth/.

3. For example, theologian William Lane Craig has argued that all humans must come from a single human couple, Adam and Eve. Because there is strong genetic evidence that the population of *Homo sapiens* was never just two, Craig pushes that original human couple back more than 500,000 years ago to *Homo heidelbergensis*. (The current genetic tools for estimating population size don't extend that far back.) That means *heidelbergensis* and Neanderthals should be called humans (and image bearers) too. See Craig's *In Quest of the Historical Adam: A Biblical and Scientific Exploration* (Grand Rapids, MI: Eerdmans, 2021).

CHAPTER 14: DEGREES AND KINDS IN THE CAVES

1. Charles Darwin, "Comparison of the Mental Powers of Man and the Lower Animals—Continued," chap. 4 of *The Descent of Man*, in *The Origin of Species and The Descent of Man* (New York: Modern Library, 1936), 494–95.

2. G. K. Chesterton, *The Everlasting Man* (New York: Dodd, Mead, 1926), 16.

CHAPTER 15: FINDING DEEPER FAITH AMONG THE REDWOODS

1. "Uniquely Unique: What Does It Mean to Be Human?," *Language of God* podcast, July 22, 2021, https://biologos.org/podcast-episodes/uniquely-unique-what-does-it-mean-to-be-human.

2. Jim Stump, "Scientific Testimonies to Human Uniqueness," BioLogos, May 21, 2018, https://biologos.org/post/scientific-testimonies-to-human-uniqueness.

3. Ernst Fehr and Urs Fischbacher, "The Nature of Human Altruism," *Nature* 425 (October 23, 2003): 785.

4. Kevin N. Laland, *Darwin's Unfinished Symphony: How Culture Made the Human Mind* (Princeton, NJ: Princeton Univ. Press, 2017), 14.

5. Raymond Tallis, *Aping Mankind* (Abingdon, UK: Routledge, 2016), 11.
6. Mary Midgley, *The Ethical Primate: Humans, Freedom, and Morality* (Abingdon, UK: Routledge, 1994), 24.
7. The first attribution of this I have found is Richard D. Alexander, *How Did Humans Evolve? Reflections on the Uniquely Unique Species* (Ann Arbor: Univ. of Michigan Special Publication 1, 1990), https://deepblue.lib.umich.edu/bitstream/handle/2027.42/57178/SpecPub_001.pdf.

CHAPTER 16: WHAT HAPPENED TO THE SOUL?

1. See Jeffrey M. Burns and Russell H. Swerdlow, "Right Orbitofrontal Tumor with Pedophilia Symptom and Constructional Apraxia Sign," *Archives of Neurology* 60 (March): 437–40.
2. Dennis Coon, *Essentials of Psychology*, 8th ed. (Belmont, CA: Wadsworth/Thomson Learning, 2000), 2.
3. Francis Crick, *The Astonishing Hypothesis: The Scientific Search for the Soul* (New York: Touchstone, 1995), 3.

CHAPTER 17: BONES AND RELICS

1. To view an image of a bat skeleton, see "Chiroptera: More on Morphology," UC Museum of Paleontology, https://ucmp.berkeley.edu/mammal/eutheria/chiromm.html.
2. Thomas Nagel, "What Is It Like to Be a Bat?," *Philosophical Review* 83, no. 4 (1974): 435–50.
3. Jack Hitt, *Off the Road: A Modern-Day Walk Down the Pilgrim's Route into Spain* (New York: Simon and Schuster, 2005), 27.
4. M. Fillmore and M. Vogel-Sprott, "Expected Effect of Caffeine on Motor Performance Predicts the Type of Response to Placebo," *Psychopharmacology* 106 (1992): 209–14.
5. Chris L. Kleinke, Thomas R. Peterson, and Thomas R. Rutledge, "Effects of Self-Generated Facial Expressions on Mood," *Journal of Personality and Social Psychology* 74, no. 1 (1998): 272–79.
6. Gregory, Bishop of Nyssa, "On the Making of Man," Christian Classics Ethereal Library, NPNF2-05, https://ccel.org/ccel/schaff/npnf205/npnf205.x.ii.ii.i.html.

CHAPTER 20: FINDING DEEPER FAITH WITH HELEN KELLER

1. See, for example, Ernst Cassirer: "No longer in a merely physical universe, man lives in a symbolic universe." *An Essay on Man* (New Haven, CT: Yale Univ. Press, 1944), 25.

2. Charles Foster, *Being a Human: Adventures in Forty Thousand Years of Consciousness* (New York: Metropolitan Books, 2021).

3. Michael Newton, *Savage Girls and Wild Boys: A History of Feral Children* (New York: Picador, 2002), 194–95.

4. Newton, *Savage Girls and Wild Boys*, 43.

5. Newton, *Savage Girls and Wild Boys*, 44.

6. Helen Keller, *The World I Live In and Optimism: A Collection of Essays* (Mineola, NY: Dover, 2002), 48.

7. Yuval Noah Harari, *Sapiens: A Brief History of Humankind* (New York: Harper, 2015), 24.

CHAPTER 21: HUMAN EVIL

1. Thomas Hobbes, *Leviathan* (New York: Collier Books, 1962), 100.

2. See, for example, Steven Pinker, *The Better Angels of Our Nature: A History of Violence and Humanity* (London: Penguin, 2012), 42.

3. W. D. Stansfield, "The Bell Family Legacies," *Journal of Heredity* 96, no. 1 (January/February 2005): 1–3, https://doi.org/10.1093/jhered/esi007.

4. "American Breeders Association," University of Missouri Library, Special Collections and Archives, 2011, https://library.missouri.edu/special collections/exhibits/show/controlling-heredity/mizzou/breeders.

5. David A. DeWitt, "The Dark Side of Evolution," Answers in Genesis, May 10, 2002, https://answersingenesis.org/sanctity-of-life/eugenics/the-dark-side-of-evolution/.

6. Henry M. Morris, "Evolution and Modern Racism," Institute for Creation Research, October 1, 1973, https://www.icr.org/article/evolution-modern-racism/.

7. S. Syropoulos et al., "Bigotry and the Human–Animal Divide: (Dis)belief in Human Evolution and Bigoted Attitudes Across Different Cultures," *Journal of Personality and Social Psychology* 123, no. 6 (2022): 1264–92, https://doi.org/10.1037/pspi0000391.

CHAPTER 22: NATURAL EVILS?

1. Richard Dawkins, *River Out of Eden* (New York: Basic Books, 2008), 132.

CHAPTER 23: WHAT GOD CAN'T DO

1. Stephen Jay Gould, *Wonderful Life: The Burgess Shale and the Nature of History* (New York: W. W. Norton, 1989), 289.

2. "Simon Conway Morris: Complete Imponderables," *Language of God* podcast, July 14, 2022, https://biologos.org/podcast-episodes/simon -conway-morris-complete-imponderables.

3. That's what he said in the podcast. In the relevant section of the book, he says, "Their combinatorics can yield staggeringly large numbers of potential biological alternatives, where 10^{200} (or even more) is typical." Simon Conway Morris, *From Extraterrestrials to Animal Minds: Six Myths of Evolution* (West Conshohocken, PA: Templeton Press, 2022), 49–50.

4. Conway Morris, *From Extraterrestrials to Animal Minds*, 51.

5. Simon Conway Morris, *Life's Solution: Inevitable Humans in a Lonely Universe* (Cambridge: Cambridge Univ. Press, 2003), xii.

6. "Simon Conway Morris: Complete Imponderables."

CHAPTER 24: BUILDING MORALITY

1. "Uniquely Unique: Morality, Language, Culture," *Language of God* podcast, August 5, 2021, https://biologos.org/podcast-episodes/uniquely -unique-morality-language-culture.

2. Sarah Brosnan, "Why Monkeys (and Humans) Are Wired for Fairness," TED, December 2020, https://www.ted.com/talks/sarah_brosnan_why _monkeys_and_humans_are_wired_for_fairness.

3. Penny Spikins, *How Compassion Made Us Human: The Evolutionary Origins of Tenderness, Trust, and Morality* (Barnsley, UK: Pen & Sword Books, 2015), 67–68.

4. Spikins, *How Compassion Made Us Human*, 177.

5. Spikins, *How Compassion Made Us Human*, 75–76.

6. David Graeber and David Wengrow, *The Dawn of Everything: A New History of Humanity* (New York: Farrar, Straus and Giroux, 2021), 14.

7. Graeber and Wengrow, *The Dawn of Everything*, 20.

CHAPTER 25: EVOLVING IMAGE BEARERS

1. Paul Bloom, *The Sweet Spot: The Pleasures of Suffering and the Search for Meaning* (New York: HarperCollins, 2021), 43.

2. Bloom, *The Sweet Spot*, 192–93.

3. Rebecca Solnit, *A Paradise Built in Hell: The Extraordinary Communities That Arise in Disaster* (New York: Penguin, 2020), 6.

4. Bloom, *The Sweet Spot*, 35.

CHAPTER 26: FINDING DEEPER FAITH WITH SIMONE WEIL

1. Simone Weil, "Some Thoughts on the Love of God," in *On Science, Necessity, and the Love of God*, trans. and ed. Richard Rees (London: Oxford Univ. Press, 1968), 153–54.

2. Simone Weil, *Gravity and Grace* (London: Routledge, 2002), 81.

3. Simone Weil, "The Love of God and Affliction," in *On Science, Necessity, and the Love of God*, 193–94.

CONCLUSION

1. Robert J. Russell, "Eschatology in Science and Theology," in *The Blackwell Companion to Science and Christianity*, ed. J. B. Stump and Alan G. Padgett (Malden, MA: Wiley-Blackwell, 2012), 545.

2. Holmes Rolston III may have coined this phrase in his book *Three Big Bangs: Matter-Energy, Life, Mind* (New York: Columbia Univ. Press, 2010).

3. "Bill Newsome: Neuroscience, Faith, and Free Will," *Language of God* podcast, May 27, 2021, https://biologos.org/podcast-episodes/bill-newsome-neuroscience-faith-free-will.